ORIGINAL
AUSTIN SEVEN

Other titles available in the *Original* series are:

Original AC Ace & Cobra
by Rinsey Mills
Original Aston Martin DB4/5/6
by Robert Edwards
Original Austin-Healey (100 & 3000)
by Anders Ditlev Clausager
Original Citroën DS
by John Reynolds with Jan de Lange
Original Jaguar XK
by Philip Porter
Original Jaguar E-Type
by Philip Porter
Original Jaguar Mark I/II
by Nigel Thorley
Original Land-Rover Series I
by James Taylor
Original Mercedes SL
by Laurence Meredith
Original MG T Series
by Anders Ditlev Clausager
Original MGA
by Anders Ditlev Clausager
Original MGB
by Anders Ditlev Clausager
Original Mini Cooper and Cooper S
by John Parnell
Original Morgan
by John Worrall and Liz Turner
Original Morris Minor
by Ray Newell
Original Porsche 356
by Laurence Meredith
Original Porsche 911
by Peter Morgan
Original Sprite & Midget
by Terry Horler
Original Triumph TR
by Bill Piggott
Original VW Beetle
by Laurence Meredith

ORIGINAL
AUSTIN SEVEN

by Rinsey Mills

Photography by Paul Debois and the author
Edited by Mark Hughes

FRONT COVER
The Austin Seven in one of its most popular forms – an RP saloon from 1934. This car was built just before a watershed in Seven evolution when the Ruby and associated models adopted a cowled radiator.

HALF-TITLE PAGE
Although not quite as pure-bred as the famous Ulster, the sports models that followed it during the 1930s – this is a 1936 Nippy – are delightful and much sought-after today.

TITLE PAGE
Styling changes for the Ruby transformed the Seven's appearance. This 1937 car has the more sloping windscreen and different window surrounds that had been introduced a year earlier.

BACK COVER
The Ulster: this is a supercharged model dating from 1930, seen against an appropriate backdrop of tuning sheds at Brooklands.

Published 1996 by Bay View Books Ltd
The Red House, 25-26 Bridgeland Street
Bideford, Devon EX39 2PZ, UK

© Copyright 1996 by Rinsey Mills
Type and design by Chris Fayers & Sarah Ward
Sub-editing by Paul Hardiman

ISBN 1 870979 68 0
Printed in Hong Kong
by Paramount Printing Group

CONTENTS

Introduction	6
Past & Present	7
Chassis	16
Bodywork	18
Interior Trim	46
Instruments	56
Engine	63
Cooling System	73
Exhaust System	74
Fuel System	75
Transmission	77
Wheels & Tyres	79
Rear Axle	80
Front Suspension	82
Rear Suspension	84
Steering	84
Brakes	85
Electrics & Lamps	87
Tools	93
American Austin	94
Dixi & BMW	101
Rosengart	107
Export Variations	114
Special Coachwork	116
Data Section	126

INTRODUCTION

Almost anyone born before 1950 in Great Britain, or for that matter a good few other countries, will have either owned an Austin Seven, done some courting in one, or at the very least been passenger in one. For some it was a brief affair but for others it was just the start of a lifelong love.

Such owners either enjoyed Austin Sevens to the exclusion of almost any other motor, or kept one at the back of the garage while more exotic and expensive machinery was played with, until the day when, Phoenix-like, the faithful little car is resurrected to relive youthful pleasures. Intended to introduce affordable and practical motoring to the general public, the Austin Seven can still provide this 75 years after its inception.

Affordable? A Seven in good condition is still a reasonable bet if you compare its value with the cost of a new car, with the possible exception of a few of the excrescent offerings from somewhere east of longitude 0°. And it is unlikely to suffer the crashing depreciation of any of them. Practical? Perfectly so, providing it is not your wish to forever travel the main arterial routes of the world at the greatest velocity you can achieve without attracting a speeding fine, while you relax on crushed velour in air-conditioned, stereophonic claustrophobia, cocooned in some latter-day Dan Daremobile.

The term 'environmentally friendly' had yet to be dreamt up in the 1920s, but the Seven scores quite well in that department because it is economical on fuel and small in size. Recyclability is not an issue because surviving complete examples are unlikely to meet that fate, and in any case the good quality materials from which a Seven is made last rather well. Besides which worshippers at the grail of the Austin Seven have been recycling these cars for years with all the incomplete or rusty cars and spare parts that still come to light.

Why then does everybody not have one? Because, quite apart from the fact that it would be something akin to feeding the five thousand if they did, the thoroughly modern motorist would not put up with the small, feeble, slow and wheezing Seven that possesses not an ounce of all-important macho appeal. The fact that the cost of buying, running and mending a Seven is tiny compared with their favourite four-wheel drive toys or people carriers would be of no interest. They might even be able to mend it themselves if they knew which end to hold a spanner. And besides this, motorway speeding fines, road rage and ram-raiding would be non-existent if everyone was equipped with a Seven!

All is not lost, however, and I would like to thank many people – Austin Seven owners almost all – for the assistance they have given with this book.

Barry Clarke, restorer and campaigner of some of the more exotic varieties of Austin Seven for more years than I care to remember, and until recently president of the Vintage Sports Car Club, kindly offered to read this manuscript. I was expecting the worst but it came back with surprisingly few corrections and what I took to be a nod of approval. The Austineers bore my 'phone calls and visits to their restoration workshops manfully and pointed me in the right direction to some of the cars to be photographed. Keith Roach, fabricator of floor pans and almost anything else you could mention, gave me more than a little guidance. Apart from his infectious enthusiasm, Colin Humphries gave me free rein with his reference books. Henry Harris offered advice, fun with his Ulster replica, and general wisdom. Hazel Gore provided contacts and guidance on the research and photographic sides. Anders Ditlev Clausager, the odd man out in not being an Austin Seven owner, threw open the Austin Seven records at the British Motor Industry Heritage Trust, of which he is archivist; sadly they are not as complete as he or I would wish. Peter Zollner and Rita Strothjohann, of BMW, which to my amusement during the preparation of this book bought the remains of the company that had started it off in cars, kindly loaned photographs free of charge. Shotaro Kobayashi, respected editor of *Car Graphic,* confirmed what I had deduced about the Seven in Japan and loaned photographs. Ron Mitchell and Dick Beagle provided information on the American Austin and Bantam.

Without any cars, however, the book could not exist, so my thanks must also go to the kind and patient people who put up with myself, Paul Debois and James Mann while we photographed their possessions. They are listed in the panel below.

Finally, to anyone embarking upon the restoration or ownership of an Austin Seven and who possesses no other literature on the subject, I would recommend that they seek out, either in original or reprint form, the correct instruction book and separate illustrated parts list for their particular car, although some of the illustrations in the latter are perversely inaccurate. And perhaps most useful and fascinating of all are the gathered extracts from the *Austin Service Journal*, volumes 1-13, 1927-39, covering the 7hp models.

OWNERS OF FEATURED CARS

Nick Turley, who owns three of the cars featured (1923 AB tourer, 1934 RP saloon and 1934 Speedy); Neill Bruce, who provided his own splendid pictures (1930 AE tourer); Roger Spearman (1935 AAL tourer); Jennifer Thorne (1936 AC Pearl); Adli Halabi (1928 Gordon England Cup); James Anderson for extricating one car from the Brooklands Museum for the day and also providing another (1930 Ulster and 1937 ARR Ruby); David Howe for driving over to Brooklands (1926 R saloon); Brian Davis for spending the day driving to various locations in the Welsh valleys (1930 RK saloon and 1934 RP saloon); Jeff Parker (1927 AD tourer); David Williams (1930 American Austin coupe); Shotaro Kobayashi (1929 type B coupé); John Clark (1936 Nippy); Mike Norris-Hill (1930 two-seater); Brian Sutton (1938 Opal); Trevor Turpin (1933 two-seater); Barry Gundry (1924 AC tourer); Chris Keevil (1933 65); Ian Hodgson (1932 Rosengart LR4); Graham Horder (1933 BMW DA-4); Keith Buckett, who gave me a nostalgic tour of the Isle of Wight (1932 Swallow saloon); Alan Garside (1930 AF tourer); and David Blunt (1937 ACA Pearl).

PAST & PRESENT

Born in Buckinghamshire and brought up in Yorkshire, the young Herbert Austin, after a grammar school education, emigrated to Australia at the age of 17. Once there he started work with an engineering firm, R Parks and Company, under the watchful eye of his uncle (they had met when the uncle visited England) who managed the firm.

Four years later, in 1887, he married a Scottish girl, Helen Dron, and soon after this, following a meeting at work, he struck up a relationship with another immigrant, an Irishman called Wolseley. Wolseley had been in the business of sheep farming and to save time and money on shearing he had set up a firm to manufacture machinery to that end.

Austin's inventive engineering mind was just what Wolseley needed and soon he persuaded Austin to join his company. Unfortunately things did not go well, mainly due to poor quality parts made by outside contractors, and in the early 1890s the company began to fold. There was, however, an English side to the business; in 1893 Austin was asked if he would like to return to his native soil to manage it, and this he accepted.

To avoid the problems encountered in Australia by sub-contracting manufacturing processes, Austin, with the approval of the directors, found and acquired a factory in Birmingham so that most of the work could take place under one roof; proper quality control would ensure a reliable product. The business began to prosper and it was not long before the firm outgrew these first premises, so a move was made to a larger factory at Aston. Other work could then be undertaken, including manufacture of the newly fashionable safety bicycle. This in turn led, as with many other cycle firms at that time, to thoughts of mechanical propulsion.

Nick Turley's unique AB tourer, first registered on 18 October 1923, with chassis/car number A1-2026 and engine number 2069. Like a beautifully patinated piece of antique furniture with its original paint and weather equipment, it was stored by the widow of the original owner from the late 1920s until it was purchased nearly 30 years later by a garage that used it for display and, with foresight, did not succumb to the temptation to restore it.

Herbert Austin's first attempt, in 1895, was a three-wheeler of Leon Bollée type which he supposedly made at home, but it is more likely that he used the engineering facilities at the Wolseley factory. The company directors became sufficiently interested to encourage further prototypes, the success of which led to the Wolseley company becoming a motor car manufacturer, in 1899. This development was further helped by the involvement of the arms firm Vickers, which gave considerable financial assistance, and in 1901 the motor car side came under the control of Vickers, becoming the Wolseley Tool and Motor Car Company Ltd.

Not unnaturally, Austin was asked to manage the running of things and cars made when he was in charge normally featured horizontal engines of capacities ranging from a single cylinder up to a 'four' of just over 5 litres. He was also an enthusiastic builder of racing machines, entering cars for the Paris-Vienna race of 1902. The best performance by a British car was made in the 1904 and 1905 Gordon Bennett races by the 11.9-litre Wolseley 'beetles', all with horizontal engines.

In spite of this the board felt it was time to move on and adopt vertical engines and shaft drive, neither of which Austin was too keen on. In any case he was ready to make his move and resigned. For some time he had nurtured the idea of setting up as a manufacturer in his own right and now he took the plunge.

With the aid of finance from people he had come to know well in the previous few years, he purchased a redundant printing works south of Birmingham. His benefactors were principally the Kayser Ellison Steel Company and Harvey du Cros, whose family owned Dunlop, du Cros himself owning the Swift Company, another cycle-turned-car manufacturer.

Production began as soon as possible with reasonably large sidevalve cars; perversely, Austin now abandoned horizontal engines, although he retained chain drive for a while before changing to shaft. From 1908 the firm produced some six-cylinder machines, including racers for that year's French Grand Prix, and in 1909 a small single-cylinder car in conjunction with Swift; in Austin form, this was called the Seven. A further useful contract, supplied by the du Cros family, was for the manufacture of an English version of the French Gladiator car, the maker of which they had an interest in. Another French car of theirs, the Clement, was concurrently built in England by Swift.

Austin cars became known for sturdy and dependable, if unexciting, construction, and the factory began to broaden its horizons with the manufacture of stationary and marine engines, including a V12. Two of these were used in the hydroplane Maple Leaf IV, the first boat to exceed 50 knots and winner of the Harmsworth Trophy in 1912-13. Commercial vehicles were also now in the repertoire, and the firm had just gone public by the start of the First World War.

In common with so many industrial concerns such as Rosengart in France and BMW in Germany, both later to become players in the Austin game, the Austin Motor Co profited enormously and grew huge during the terrible years of the war. Whether this was for philanthropic reasons or pure greed, each mixed with a degree of patriotism, would be hard to discover, but money alone could do little to assuage the loss of Austin's only son, Vernon, to a single sniper's bullet in France during 1915. Always hard-working, Austin redoubled his efforts from there on and by the end of the war the factory could boast an enormous contribution to the war effort, ranging from small gun parts to whole aircraft. For his efforts toward the cause Herbert Austin had been knighted, the company had purchased two neighbouring factories and the workforce numbered some 20,000. In 1918 he was elected a Member of Parliament but took very little part in parliamentary proceedings.

A one-model policy was adopted with the introduction of the Austin Twenty in 1919 and during the first post-war years this proved adequate, its typically robust Austin qualities finding a ready market in the Antipodes as well as at home. There was, however, a large workforce to keep in work and paid. By 1921 Austin was in trouble and an administrative receiver was appointed. The introduction of the Twelve Four in 1922, however, helped to save his bacon and a year later the company was back on the road to solvency.

During these troubled times, Sir Herbert's thoughts once again turned to the concept of a small, cheap and economical car for the masses. Legend has it that some of his first tentative design sketches were made, full size, using the billiard table at his home, Lickey Grange, as a drawing board.

His deliberations on this theme drew no support from the company's board of directors so he decided to continue work with what was to be his baby on his own account at home. To this end he secured the services of a young but extremely talented draughtsman by the name of Stanley Edge, who had first begun war work in the Austin drawing office during 1917 at the age of 14 but had been transferred to the motor car section when hostilities ceased. Sir Herbert had his eye on the young Edge and during 1921 arranged for him to work directly for him on the small car project, at Lickey Grange.

At first Austin himself was rather keen on a vehicle with some similarity to the Rover Eight, which had an air-cooled, horizontally-opposed, twin-cylinder engine, but finally, aided and abetted by Edge and very likely influenced by others, he came up with what was to be, with surprisingly few alterations, the Austin Seven.

Austin was a long-time admirer of Henry Ford,

PAST & PRESENT

The first Seven saloon, designated R, arrived in the summer of 1926. This wonderfully original car was built a few months later and first registered on 31 December. Owned by David Howe, it has chassis number 27555, engine number 27840, car number A3-6739 and body number 540.

production of whose people's car, the Model T, had long since passed the million mark. Another possible American influence was a Gray truck, left over from the war, which was used around the Austin works. It had a triangulated chassis frame with front suspension by transverse leaf spring. Over in France Peugeot's successor to the Bébé, the 668cc *quadrilette*, was obviously a cut above many other small cars with its diminutive four-cylinder, water-cooled, L-head engine in a chassis with transverse front spring and quarter-elliptics at the rear. Whatever inspired Austin, he got it basically right, as can be seen by comparing the Seven with some of the amusing but frankly abominable contraptions that were manufactured as 'cyclecars' during the post-war period.

For several months Edge worked on the full set of plans for the new car. Some of the components, due to his unfamiliarity with their design, he ingeniously scaled down from those of the larger products made by the company, but at the same time altered them to be more suitable for a small car. There were also many features only to be found on what was to become the Seven.

Although the board of the company was still not in favour of Sir Herbert's scheme, it was fully aware that he had patented many of the car's features already and, rather than run the risk of it going elsewhere should it prove successful, they agreed to the construction of prototypes at the factory. By Whitsun the first was running and by the summer a second was up and driving, having needed few alterations. Austin, impatient to get on with things and reluctant to wait for the Motor Show, arranged for the official launch of the new car to take place at Claridges Hotel, London, on July 21.

Austin had put a good deal of his own money, not to mention time, into the project and as a personal reward, in the event of a good reception by the public, he had negotiated a royalty for himself of two guineas for every car sold. In the event not only did he do well financially, but the Seven, more than any

Another saloon, this time an RK that was laid up between 1937-74 and is now owned by Brian Davis. First registered on 19 June 1930, it has chassis number 111814, car number B1-4687 and engine number 112895. Clearly visible are the roof and windscreen peak details – there were variations in the design of these features around this time.

other car made at Longbridge during that period, put the company on a firm foundation for many years to come.

Some clever or innovative products are killed off at birth due to lack of development or shoddy workmanship, but the Seven, although it did suffer minor teething troubles, succeeded where others failed or were merely tolerated. The factory, fresh from its vast expertise for rapid development acquired during the recent war, was well equipped to get the car ready for production. As for workmanship, Austin himself, no doubt mindful of his early experiences with Wolseley, had always insisted that his workforce cut no corners and used good quality materials.

Neither manufacture nor sales took off with a rush, but by the end of the first full year of production, in the autumn of 1923, around 2000 cars had been delivered. The Seven had also been tried out with a good measure of success in competition, which was the responsibility of Sir Herbert's son-in-law, Captain Arthur Waite. An Australian by birth, Waite had served with distinction in the war and had been awarded the Military Cross; in 1918 he had married the Austins' daughter, Irene, and shortly afterwards joined the company. He was an adventurous type who sought more than a desk to sit behind and, harking back to Herbert's racing days with Wolseley and later Austin, he did not have too much difficulty in persuading his father-in-law that a little competition work might bring good publicity. Before the Seven appeared he had tried, with limited success, to make some sort of racing car out of the Twenty, so the tiny new car and its possibilities for *voiturette* racing would have seemed like a breath of fresh air.

By the spring of 1923 a car had been prepared, with a pointed tail body following aircraft construction of painted fabric over wooden stringers. To cope with higher gearing, the engine was mildly tuned with a high-lift camshaft, special inlet and exhaust manifolds, and twin Cox Atmos updraught carburettors. After giving the car its baptism of fire at the Brooklands Easter Small Car Handicap as 'limit man' and managing to hold onto the lead of this short race until the end, Waite and Alf Depper, the racing department foreman, entered the same car in the Cyclecar Grand Prix at Monza, where they were rewarded with first place in the 750cc class. In August a team of three cars was taken over to France for the racing at Boulogne, but this time fortune did not shine on them: two cars were eliminated by engine bearing failure while the third crashed.

Undismayed the gallant band, normally dressed for racing in white overalls with the Austin script emblazoned across their backs and matching linen helmets, continued racing the same cars in England during 1924. Another foray to the continent brought them a second and third in class at the

PAST & PRESENT

The car that so many Austin Seven enthusiasts aspire to – the EA Sports, universally known as the 'Ulster'. Owned by Dr James Anderson, this 1930 supercharged car was raced for its first few years by a gentleman called Seyd; chassis number is 115878 and engine number is 107995.

Voiturette Grand Prix on the Sarthe circuit. By now Gordon England had built a single-seater Seven for use at Brooklands, the start of his association with the company, and a few private owners were beginning to use their Chummies for reliability trials and other gentler events – all good publicity for Sir Herbert's baby.

Pre-release publicity material had indicated that the price of the new car would be £225, but in the event it cost £165 ex-works when production really got under way in 1923. This figure was reduced to £155 the following year, to £149 in 1925, and to £145 by the autumn of 1926. These prices were for the Chummy as the factory had not started to make its own saloon until 1926; the cheapest version of this was £150, complete with Triplex glass.

By that time Sir Herbert must have realised that his 'motor for the millions' was unlikely to become literally that as it was only towards the end of 1926, after four years of production, that the 25,000 mark was passed. A couple of years later Ford was to begin the production of his Model A which, in contrast, was to last a similar period and total around 5 million. Poor old Sir Herbert was just not in the same league, but nevertheless the company, largely thanks to the Seven, was doing nicely and the little car was being exported to other parts of the world, although it was not until the following year, 1927, that Austin began to grant licences for versions of the Seven to be manufactured abroad, starting with Dixi in Germany.

Output in 1927 took a leap and almost equalled the number made in the previous four years, with the 50,000th car leaving the factory at the end of November. This rate of expansion could not be maintained, but 100,000 Sevens were on the road by the end of the decade, the factory-bodied cars supplemented by the many special-bodied versions that were available by that time.

The factory had from time to time, since Waite's first racing Sevens, experimented with various special versions, including a supercharged single-seater made for Waite in 1925. However, towards the end of 1927 work began on a car which, although to be supercharged, was intended to be recognisable to the general public as a sports Austin Seven rather than an out-and-out racing car. The first of these five 'Super Sports' built was sent out to Australia and the redoubtable Waite, who had returned there to work at the distributor, managed to win outright the 100-mile 'Grand Prix' organised by the Victorian Light Car Club at Philip Island in March 1928 – and he even managed to lap faster than a couple of 1½-litre Bugattis! All other places in the 750cc class were taken by privately-entered Austin Sevens – a foregone conclusion as they were the only entrants of that capacity. The other Super Sports remained in England and were used in competition for a while.

11

Although they were the ancestors of the Ulster, they had not yet developed fully, having among other things straight rather than dropped front axles.

During the same year Lucien Rosengart began to produce his version of the Seven in France, while back at home Gordon England, Mulliner and Swallow were all in full production of their Seven bodies. The following year, 1929, saw the entry of BMW into the car market when it took over from Dixi as producer of the licensed Austin Seven in Germany. That year the Ulster Tourist Trophy was won by Caracciola in a 38/220 Mercedes with the Austin works Ulsters in third and fourth places. This was a handicap event but even so the Sevens were lapping at over 60mph. The Ulster, or more correctly the Sports, became available to the general public at the beginning of 1930 priced at £225 and £185 for the supercharged and unsupercharged versions.

In October 1930 the Motor Show held at Olympia, which had been enlarged in time for this 24th show, saw several Sevens on the Austin stand. Both front and rear brakes were now coupled to the foot pedal and there was a new-look, higher radiator. Prices for this watershed year were £94 10s for a chassis, £122 10s for the tourer and £130 for either fabric or coachbuilt saloon with a sunroof thrown in for an extra £5. Austin's show publicity made much of the New Seven and invited people to, 'Examine the new car. Note that it is a REAL four-seater. That improvements have been made to the body AND to the chassis – the new brake gear, the stiffer crankshaft, the reserve petrol supply and many other details'. On the coachwork stands there were more Sevens, including a Swallow version sporting a redesigned radiator. And if the Seven owner could not run to a new car, then the numerous accessory manufacturers surely had something to tempt him.

The Ulster had mixed fortunes in 1930. The cars were beaten to the team prize by the MG Midgets in the Double Twelve at Brooklands, managed third and fifth places in the Irish GP, and only fifth in the Ulster TT – but they ended the season with a quite brilliant win in the 500-mile race at Brooklands.

The depression had been hard and finished off many companies but Austin had ridden it out, the Seven selling as steadily as ever and seeming set to continue into the next decade with a few simple improvements. Besides that, the third of Sir Herbert's licensing agreements had borne fruit and the American Austin had finally gone into production that summer.

As the 1930s progressed the Seven began to gain middle-aged avoirdupois and the factory, never really getting to grips with this problem, embarked upon small measures to retain performance of both accelerative and decelerative natures – but the Seven had lost its youthful sprightliness. This did not put off the customers; just under 200,000 cars had been delivered in total by the time the Ruby came along during the summer of 1934. The foreign versions were a little less consistent: although Rosengart was managing quite well, the American Austin enterprise lurched from one disaster to another, and BMW had by now moved on to other things.

Sports Austins continued to be made in small numbers. The Ulster was discontinued in 1932 and replaced in 1933 with the much tamer 65, which had no pretensions to be a racer. This was in turn replaced by the almost identical Nippy, which lasted until 1937; for dedicated boy racers the Speedy was made in small numbers in 1934 and 1935. Now that BMW was no longer involved, Austin itself also actively entered the German market once more and the 65, Nippy and Speedy were all available in left-hand drive form marketed by Willys-Overland, Crossley GmbH Berlin. Sir Herbert Austin visited this company's stand at the Berlin Motor Show in 1935 and had the dubious pleasure of meeting the Führer, who is on record as having told him, through an interpreter, that the Seven had a special place in his heart due to a Dixi having been his first car!

The year of the greatest Seven sales, just over 27,000, also came in 1935, but then they gradually tailed off. Even with the more modern looks introduced with the Ruby series, there was now much stiffer competition from, among others, Morris and Ford, whose products offered more than the Austin. Quite apart from its looks, the Morris Eight had hydraulic brakes and a greater degree of sophistication. Quite what the Ford Eight had, apart from a propensity to wander all over the road, defeats me but perhaps it was the price – very cheap.

The most radical mechanical changes to the Seven during the 1930s took place in 1932 when an extra ratio was added to the gearbox, and in 1936 when a central main bearing was added to the engine. The former helped the now heavier cars gain momentum and negotiate hills, although the extra bearing provided no performance increase.

Competition activities continued during this period but now took two distinct directions: on the one hand there were cars developed from production components and on the other were single-seaters designed specifically for sprints, hillclimbs and races. It was realised, by the close of the 1930 racing season, that to continue to race sports cars that were recognisable as production Sevens was an unattractive proposition due to the improving performance of the overhead camshaft MG Midgets in the 750cc class: it was therefore decided to build a single-seater which would attempt to be the first 750cc car to attain 100mph. Unfortunately by the time the car was anywhere near ready MG had achieved this goal at the Montlhéry track in France.

The Austin's highly stressed supercharged engine gave months of teething troubles but by the end of

PAST & PRESENT

Rarity makes the Speedy, of which only around 60 are thought to have been built, enormously desirable today, even if it is not such a pure 'racer' as the Ulster. Owned by Nick Turley, this example was manufactured in June 1934; chassis number is 196481 and engine number is 198894.

the summer it managed to last long enough to take a few records at Brooklands and in the process became the first 100mph 750cc car in England. This car was joined for the 1931 500-mile race by a team of three newly-built single-seaters which were soon nicknamed 'Rubber Ducks' on account of their looks. All four cars retired with overheating, but they were soon made reliable and continued to be raced for several more seasons. But these supercharged sidevalve cars, giving a reliable 56bhp at 6000rpm, reached a pinnacle of development and Austin decided to put in hand its ultimate sidevalve car.

The special engine was designed by Murray Jamieson, a newcomer to the factory experimental department. Apart from sharing the bore and stroke as well as a few other dimensions, this was a completely new unit. With a supercharger running at twice engine speed it produced a reliable 70bhp at 7000rpm with more available for short periods. In the middle of the ding-dong battle with MG for Class H records it managed over 122mph in 1934 when fitted with a streamlined body, but by then MG was up to 128mph. Subsequently it was rebuilt as a racing car and achieved class wins at Brooklands and Shelsley Walsh, these successes prompting the factory to build a second car to be sent to Germany for use in Europe.

Austin was anxious that his company should continue to gain the publicity which racing attracted and so in 1934, conscious that the sidevalve layout had truly run its course, he asked Jamieson to design a completely fresh 750cc racing car. This was financed by Austin himself, the only stricture being that the suspension should have the same basic layout as the Seven so that there was at least a tenuous link. The fruits of this were the three twin-camshaft single-seaters, the short-stroke engines of which were variously reported to give an easy 90bhp for distance work with up to nearly 120bhp on special fuel mixes at over 7500rpm.

A series of mechanical teething troubles beset these cars and one driver ended up in hospital with the car wrecked. As a result there was talk of aborting the whole project and Jamieson decided his talents were better employed by ERA. Eventually reliability was found and the cars enjoyed a successful 1937 season. The Coronation Trophy at Donington was a high point with all the races of the day falling to Austin, but no further serious development was done on the design and the cars were used in the same form until the war.

The alternative course taken by the competition department involved the preparation of cars for reliability trials and long-distance events. The Speedy was used as a basis for this and although the standard

car was a veritable sheep in wolf's clothing it was successfully modified for competition.

For the 1935 trials season a team of three was prepared with many of the basic Speedy ingredients, although built on Ruby chassis because this allowed the now almost obligatory slab tank and twin spares to be used without the rear of the car falling to pieces.

The Pearl cabriolet (above) was a charming variation on the Ruby theme. This is the first type, the AC Pearl, and is owned by Jennifer Thorne. First registered on 2 September 1936, it has chassis number 249785 and engine number 250772. Registered in Exeter, this Opal (left) was discovered, in reasonably complete and original condition, on nearby Exmoor some years ago.

PAST & PRESENT

In the 1950s and '60s the long-suffering Seven provided many a young man with unusual motoring experiences. This car (above), a 1925 AC tourer, was modified with various components from later Sevens and other sources. While conveying its slightly sozzled owners back from a pub in South Wales one night, it decided it had had enough of these indignities and with a hop, skip and a jump ended up like this. All parties survived and the car has subsequently been renovated. Even today cars still await discovery such as this 1927 AD Tourer (right). A cow suffering from milk fever collapsed on it some years ago and now, in a bed of hay, the little car gathers dust and rust.

The engine was also rather a special variant which gave 27bhp.

As well as the trials cars, the factory rather ambitiously built four more cars for the 1935 Le Mans 24 hour race – three of them with slab tanks and one with the pointed tail of the Speedy. Their target was to qualify for the Biennial Cup, and this they managed; two of the cars retired but two finished at the back of the field which was sufficient. Another batch of cars was built, with the new three-bearing engine, for the 1936 race but this was cancelled due to industrial action in France. In 1937 the team again tried its luck at the Biennial Cup, but it was not to be theirs as all the cars retired with centre main bearing failure.

They partially exonerated themselves with second and third places in their class at Donington in a 12-hour sports car race, but thereafter they, like the 1935 versions, were converted for trials use. By now the trials cars, or 'Grasshoppers' as they had been nicknamed, were running with belt-driven Centric superchargers as well as raised suspension and had become a force to be reckoned with due to their light weight and manoeuvrability. In all 12 were built and continued to do very well in trials right up until the war.

The production Seven, meanwhile, had been quietly dropped from Austin's range. The last saloon, after just over 290,000 cars and chassis had been produced, left the factory in January 1939, followed shortly afterwards by the last van. Austin's replacement for the Seven, the Big Seven and then the Eight, good cars though they may have been, never achieved the same cachet. Not until two decades later, when Austin had merged with Morris and engulfed others to form the British Motor Corporation, would the company once again produce a small car to catch the public's imagination – the Mini. It was perhaps no coincidence that the Austin version of the Mini was named the Seven.

Austin Sevens never lost their appeal even in creaky and arthritic old age. The sense of affection and fun they engender ensured that sufficient numbers survive for us to enjoy today. As the years passed many fell by the wayside, but some still lie in wait even now, for someone to discover them and bring them back to life. Others underwent strange metamorphoses as they were pressed into service for student transport, club racers and other devices which almost defy description. A very few have been kept, largely unmodified, either in careful use or stored virtually unused, and it is these rare survivors which best tell us the history of Seven.

Even now, not all Austin Seven owners are particularly interested in the historical side of things nor are they over-bothered with originality – and why should they be? And a good number of people are already fully aware of exactly how the cars were put together in the first place, and I do not presume that this book is for them. Somewhere between the two factions lies the growing band of enthusiasts who would like to ensure that the Austin Seven which they own is in approximately the same specification as it left the factory or would like to undertake the restoration of one. The information contained here may be of some use to them.

15

CHASSIS

When Sir Herbert Austin designed the chassis frame for his baby car he cleverly distilled aspects from various other manufacturers, not least from Henry Ford, into his own unique product.

The channel sections were made from very high grade steel of 12 gauge and in plan view take the form of an A, the top closed by a steel forging that carries the front spring. In addition there are a pair of crossmembers of the same 12 gauge, one in the centre and the other just short of the rear.

The very first cars had no shock absorbers so the plates which carry these at the rear of the chassis were absent until car number A1-3145. Some concern was shown that the chassis sagged a little with age and so during August 1926, from car number A1-3361, the side members were made a little deeper to counteract any tendency for this to happen.

By 1927 the factory had to own up to the fact that the chassis gave inadequate support to the rear of the bodywork and, although blaming owners for overloading the cars, produced pairs of pressed steel chassis extensions which were supplied free of charge to agents for them to fit – especially to saloons – if required by customers. In December these frame extensions were standardised on production chassis and at the same time the rear shock absorber brackets were abbreviated; this took place from chassis number 50901.

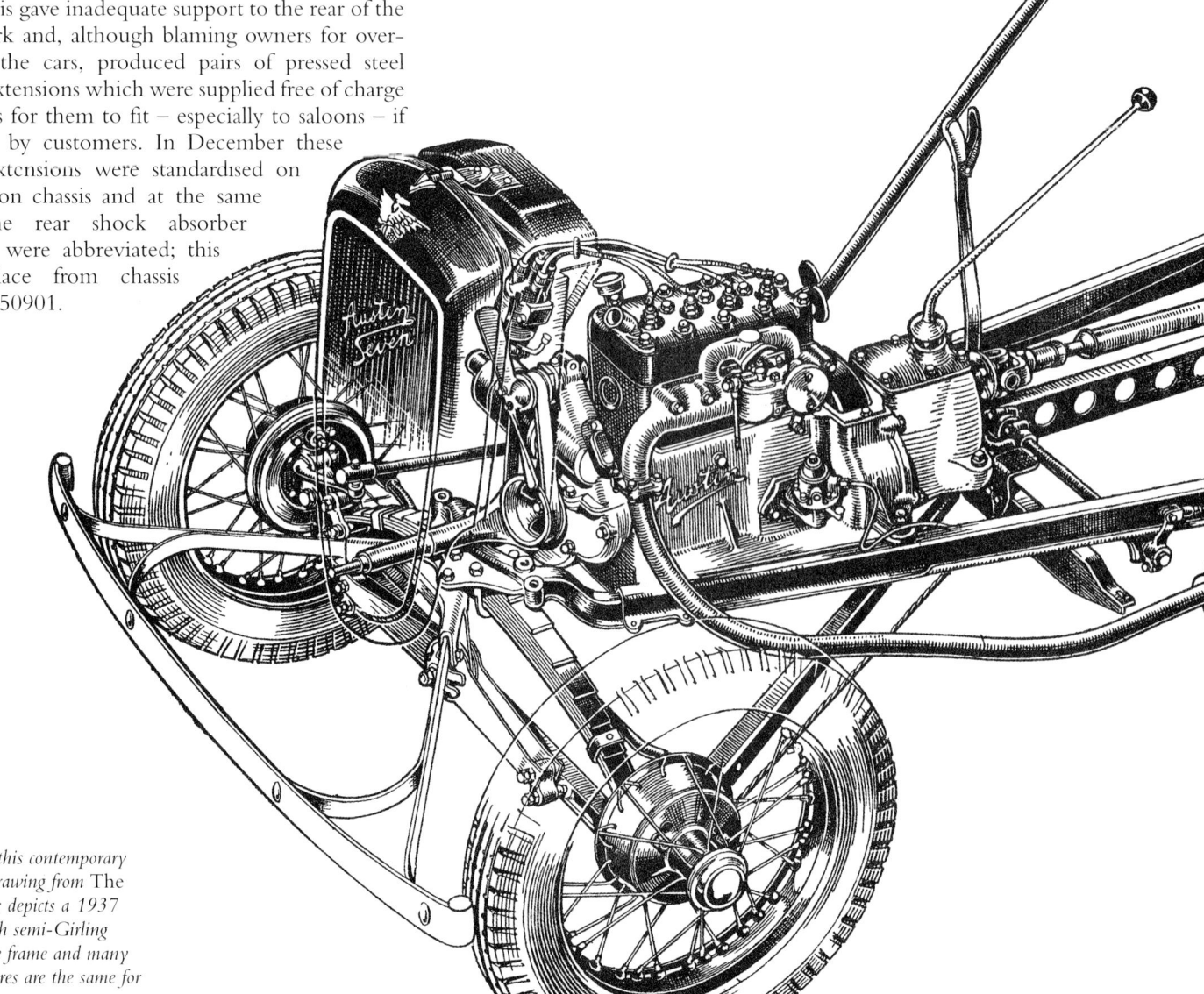

This 1929 model year chassis shows various components applicable to that year, but no attention should be given to the paintwork because a special finish was applied for exhibition purposes.

Although this contemporary cutaway drawing from The Light Car *depicts a 1937 chassis with semi-Girling brakes, the frame and many other features are the same for all low-frame cars.*

Chassis

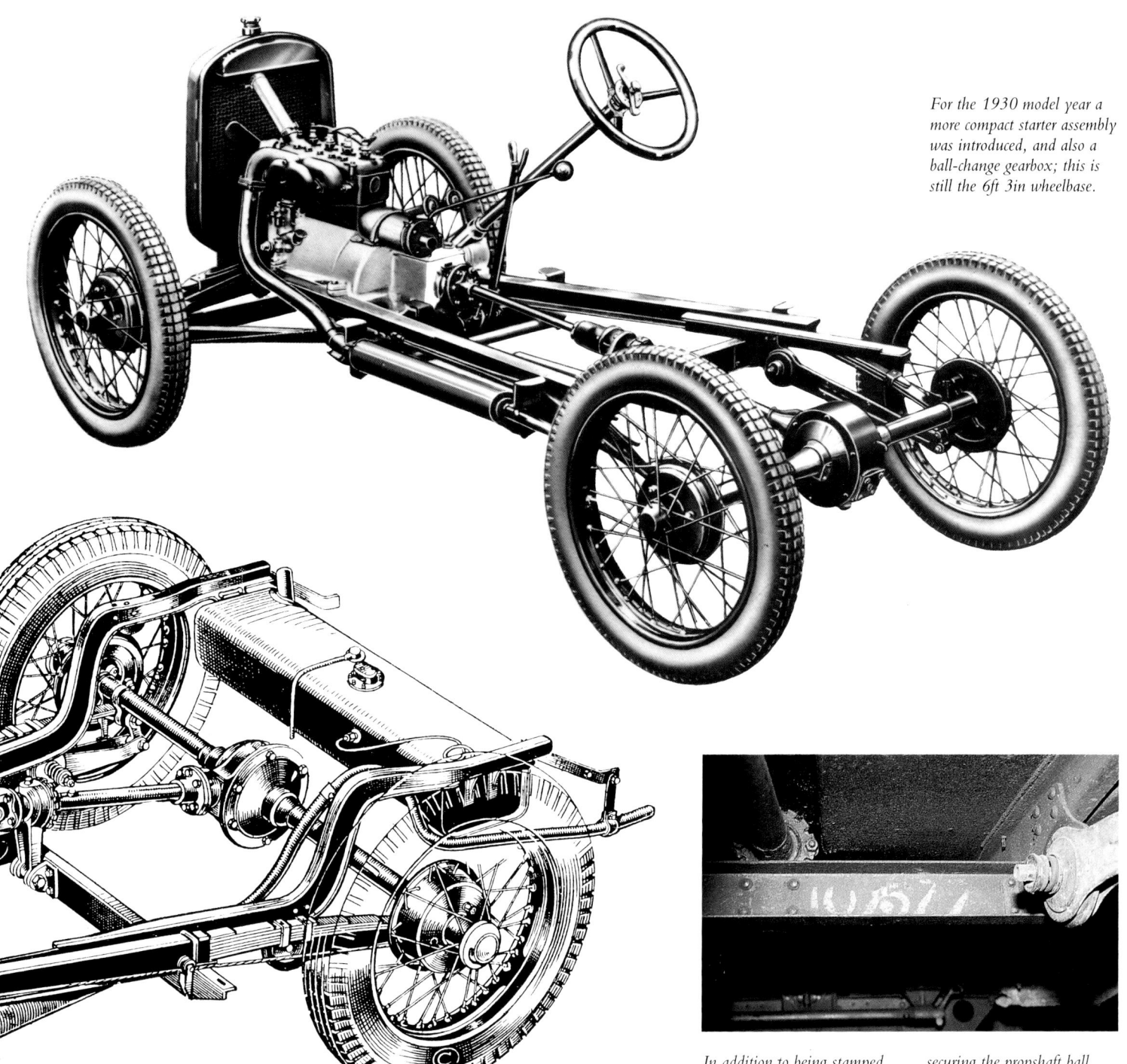

For the 1930 model year a more compact starter assembly was introduced, and also a ball-change gearbox; this is still the 6ft 3in wheelbase.

In addition to being stamped on the nearside rail between the engine mounting feet, the chassis number (107577 visible here) was stencilled on the rear crossmember. To the left of this are the four rivets securing the propshaft ball joint and the castellated adjusting nut. To the right is one rear shock absorber and on the crossmember in the background is the brake cross shaft.

With the advent of coupled brakes in 1930, around chassis number 113000, a pressed steel longitudinal member was added between the two crossmembers to serve as a mounting for the added mechanism. The factory also supplied this piece plus the other necessary parts to dealers in case customers asked them to convert older cars. A longer chassis frame, which increased the wheelbase from 75in to 81in, was introduced during the autumn of 1931.

The low-frame chassis that was to carry the Ruby body and its affiliated models began production in July 1934 at chassis number 198596. This chassis is instantly recognisable by the extensions at the rear which first sweep up then continue horizontally rearwards. Sports derivatives, with the exception of the 'Grasshopper' and Le Mans models, continued to use the previous type of chassis.

All chassis frames and their associated components were finished in black: the chassis number was normally stencilled on the rearward edge of the rear crossmember as well as stamped on the nearside chassis rail.

17

BODYWORK

Due to the design of the chassis it was decided to erect the bodywork on a fabricated sheet steel floorpan which was made in sections and riveted together. The tourer body was constructed with a light ash frame and panelled in aluminium; wings and bonnet on factory-bodied cars were always steel.

The R-type saloon, introduced in 1926, was put together in the same way but the fabric version had a framework of ash and plywood which was then clothed in canvas, padding and finally Rexine. On some fabric saloons, such as the RK of 1928, parts of the body such as the scuttle top and doors were still panelled underneath the overall fabric covering.

Dimensions of the bodywork were generally increased slightly as time went on – for example the AE tourer of 1929 was approximately 2in wider and longer than its predecessor. During the production of the RK saloon in the same year the factory began to make small alterations, seemingly without rhyme or reason, to various examples. For instance, the size and construction of the peak over the windscreen differed on some cars and the screen itself was deeper on a few. The RL saloon introduced during 1930 underwent even more small variations during its life, due mainly to the changeover in styling to the short scuttle; a louvred bonnet was also introduced during the run of this model, as it was on the contemporary tourer. Panelwork on the saloon, other than for a few fabric versions, was then of pressed steel over a reduced ash frame; tourers with steel panelwork were made from the AF of 1931.

The RN saloon, introduced in the autumn of 1931, was built on the longer chassis and was also wider; due to the longer body the doors no longer continued over the rear wheelarches. Tourers continued to be made with small doors until the AH of 1932, which was built with the larger doors now possible due to increased body length. Both the 65 and the Speedy used aluminium panelwork over an ash frame but the replacement for the former, the Nippy, had steel panels.

The Ruby and its open versions had pressed steel bodywork with even less timber in evidence; what was there was used mainly for the attachment of trim and some stiffening. Spot-welding was now used almost exclusively for assembly of floorpan and bodyshell.

As the Seven evolved and different bodies were introduced, the wings and running boards also changed to take account of both modernisation and necessary alterations to the shape.

The first pre-production prototypes had no running boards, so the front wings ended with a little flourish, but from March 1923, when production started, running boards were fitted and curled up at either end to meet the front wings, which were straight, and the rear wings. Other than on saloons and coupés, for the next six years the wings and running boards remained more or less the same, but did in fact go through several minor variations.

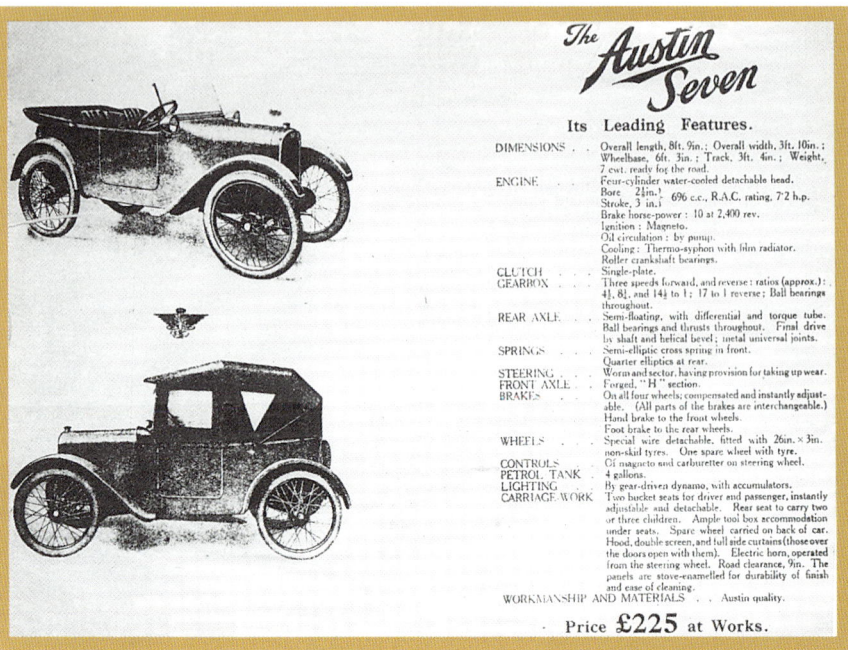

Pre-production sales brochure of 1922 shows one of the three hand-built prototypes. Note the correct absence of running boards: the single surviving car, now in the Science Museum in London, has had running boards added subsequently. The anticipated price of £225 dropped to £165 when the AB tourer went on sale.

In September 1929 a more domed profile was introduced for the front wing, first in 'short' form and then 'long'. The 'long' variety was introduced to suit altered running boards, which became flat and required the wings to curve to meet them.

At the beginning of 1930 front and rear wings with a broad swage line around their outer edge were introduced, and the general shape also became more curvaceous. Rear wings had separate variations depending on whether they were intended for long- or short-scuttle saloons, vans or coupés.

When the longer chassis cars appeared, in 1931, a much more fulsome front wing began to be used. At the same time a running board covered with embossed rubber moulding was introduced. The rear wings continued to have small variations according to body style.

The Ruby and its affiliated models had a fresh design of wing with the exception of the chromium radiator Opals, which carried on using up the stock of old wings. With the second series of Ruby type cars introduced in 1936, both front and rear wings were given a fuller skirt on the trailing edges.

In the case of a car needing new wings, there is little chance of finding 'new old stock' replacements, but several firms specialise in the manufacture of the correct type and would be able to advise the exact variety required for a particular car.

The following is a list of the various models with factory bodywork produced at Longbridge during the life of the Austin Seven, and includes some of the prototypes that never went into production. Two non-factory bodies, by Mulliner and Thomas Startin, are also described in this section.

BODYWORK

The scooped scuttle and pram hood (amazingly the original one on this car) of the AB tourer are clearly visible from the side, as are the speedo drive and cable that were accessories fitted when the car was new; this drive is similar to the later factory type, but a belt rather than a spring is used from the propeller shaft. Like so many designs that were updated over the years, the first is among the most pleasing aesthetically.

The considerable drop-away of the body aft of the sloping windscreen and vertical door sides all identify this as the AB tourer, the first production model of 1923.

Original Austin Seven

A 1924 AC tourer belonging to Barry Gundry, chassis number A1-5941 and engine number 5959. This is possibly the prettiest of the tourers, retaining the pram hood and delicacy of the earliest cars but without the scooped scuttle. Remarkably this car also retains its original hood.

Jeff Parker's 1927 AD Tourer, chassis number 43879, engine number 44239 and body number A5-698. This later version has the door handles pointing downwards and Lucas lamps. The starting handle is notched into its correct position.

BODYWORK

An early AD tourer with 6-inch brakes and CAV headlamps (above); this car would have been made in the spring of 1926. Between September 1926 and January 1927 the factory fitted the headlamps next to the radiator (above right), but due to uncertainty about legality they went back to scuttle mounting until 1928.

Tail view of 1927 AD tourer shows Lucas AT201 rear light with duplicate on the nearside to conform with the law; the rear window of the hood is a little too large and set too high.

TOURERS

XL 1, XL 2 and XL 3 These were the three hand-built prototypes constructed during the spring and early summer of 1922. They were registered OK 2950, OK 3261 and OK 3537. The latter is on display, in not exactly original form, at the Science Museum in London.

AB Family Model Chassis numbers 1–100 had the 696cc engine, and then the 747cc engine was used from chassis number 101. Aluminium body with 'scooped' scuttle, doors with vertical sides, and a sloping windscreen; no running boards at first. Hartford shock absorbers from chassis number 3145.

AC Tourer Introduced during the summer of 1924 at chassis number 4602. Vertical windscreen and rear edge of doors sloping; Austin-manufactured shock absorbers from chassis number 10653.

AD Tourer The most common vintage body appeared in February 1926 at chassis number 17074; the first batch had 6in brakes but from then on 7in. Lower section of the windscreen now curved to follow the contour of the scuttle; pram hood of the earlier models replaced by one with more angular

21

Original Austin Seven

Neill Bruce's 1930 AE tourer remains in a remarkably original state; chassis number is 107577 and engine number is 106516. Purchased new by a Mrs Guthrie from Brookwood Motors of Woking, it was supplied with this special colour and accessories. Glass and weather equipment are original; yellowed windscreen is not an early example of tinting but merely the celluloid in the centre of the sandwich of the Triplex glass discolouring with age. The bodies of the AE tourers, introduced in 1929, were a little larger than their predecessors. Wings on this car are of the second, more curved, type fitted to the AE model. Due to the car's long storage away from the light, the owner has been able to preserve the pre-war tyres.

BODYWORK

The AJ tourer, introduced in September 1932, was almost the same as the AH but had the petrol tank mounted at the rear; this factory picture dates from 1933. Door handles were mounted either in this position or horizontally, seemingly in a haphazard fashion.

The Milk Delivery version, which was based on the AF tourer. Spare wheel, number plate and rear light were all attached to the tailgate. Heaven help the driver if a full churn toppled over.

lines to improve rear seat accommodation. This bodywork was offered for several years and therefore remained current through the various minor specification alterations made during that period. In October 1926 a four-piece bonnet began to be fitted from around car number 4572; external door handles were used for the first time with this body and during 1928 they were set in a vertical position.

AE Tourer Brought out in the autumn of 1929 at around chassis number 96600, this body looked similar to the AD but in fact was approximately 2in longer and wider; in addition the scuttle had an

23

An AAK Open Road Tourer, elegantly depicted in a sales brochure from 1934.

An early AAL Open Road Tourer belonging to Roger Spearman; chassis number 227469, engine number 229475 and body number AAL 23. The car even has a period tax disc, but the owner had to fabricate trafficator brackets because the fragile original Mazak ones were hopelessly broken.

BODYWORK

In order to give plenty of headroom in the back, the hood line of the Open Road Tourer looks a little ungainly. Extra lamps added for modern lighting requirements have been tastefully done; grommets blank off the bumper mounting slots. Covered spare wheel and swept tail distinguishes AAL from earlier AAK.

opening vent on either side. Until this time all body numbers had had an A prefix, but now this was changed to a B.

AF Tourer Almost the same as the previous model but with a longer louvred bonnet, the shorter scuttle still fitted with ventilators; made from June 1930 starting at about chassis number 111000. During February the following year a steel-bodied version of this style replaced the aluminium one, having a slightly shorter bonnet with a correspondingly longer scuttle; once again some of these cars left the factory with the door handles set pointing downwards. Steel-bodied cars had no half-round beading around the door edge.

Milk Delivery This quaint little device was based on the AF tourer in either aluminium or steel form and had a side-hinged rear panel in order to load milk churns or crates of bottles; additional bracketry reinforced the body sides. Total production was probably around 100.

AH Tourer From June 1932, bodywork now pressed steel with less wood framing and floor pan sections joined by spot welding; rear panel made in three sections with two vertical beads. Large rectangular doors and at first with scuttle tank, but from chassis number 159534 in September 1932 petrol tank at rear (this version was the AJ). Cars left factory with vertically-set door handles.

AAK Open Road Tourer First of the cowled-radiator tourers, starting at chassis number 198596; bonnet now with ventilator flaps rather than louvres.

AAL Open Road Tourer Made from chassis number 226017 in July 1935; gently sloping tail with spare wheel carried under a cover.

25

Original Austin Seven

SALOONS

R The first factory-bodied saloon brought out in the early summer of 1926 at chassis number 20032. Aluminium-panelled; very first cars had 6in brakes. In September of the same year a version was introduced with fabric covering to the body above the waistline – this is sometimes called the RB. Towards the end of 1927 the factory produced a handful of fabric-covered R saloons. In a similar way to the AD tourer made during the same period, the R saloon appeared in slightly different guises and so some later cars also had side ventilators on the scuttle and/or no beading around the doors.

RK Initially this saloon, introduced around chassis number 69000, was made alongside the last of the R types but quickly superseded it. To facilitate access to the rear seats, the saloon was produced, from now on until the chassis was lengthened during 1931, with wider doors which carried on partially over the rear wheelarches. Not surprisingly, this is often known as the 'Wide Door' model, and is not to be confused with the 'Wydoor' range of bodies made by Wylder of Kew at that time (see page 125). These saloons were also produced during a period of change, so the first of them had black radiators but the majority had the plated variety. Aluminium-panelled, fabric and composition versions were produced, with a variety of differing windscreen peaks, and with or without scuttle ventilators; even the roof guttering followed no exact pattern on a few cars.

RL Another saloon with a bewildering number of variations was produced from about chassis number 110000 in the summer of 1930. Pressed steel panels were now used with a raised moulding running along the bonnet sides and waistline; a more flimsy ash frame now could be said to be mounted within the bodywork rather than the panels mounted upon it. The floor pan was altered in some respects, such as footwells, and was a little stronger to cope with more weight. At first the bonnet had plain sides and the scuttle was very short (only initially without side ventilators), but this was then lengthened slightly and bonnet louvres began to be fitted. The windscreen was rather shallower than on previous saloons, and had radiused bottom corners; side

The exquisite 1926 R saloon belonging to David Howe (above and facing page). There are sliding windows for front and rear occupants, but accommodation in the back is cramped. Seen at Brooklands, this car is highly original and correct in virtually every detail.

Two variations on the use of fabric body covering for the R saloon (facing page): 1926 car (left), with 6in brakes and CAV headlamps, has fabric only above the waistline and is sometimes called the RB; 1927 car (right), with 7in brakes and Lucas headlamps, is the full fabric version.

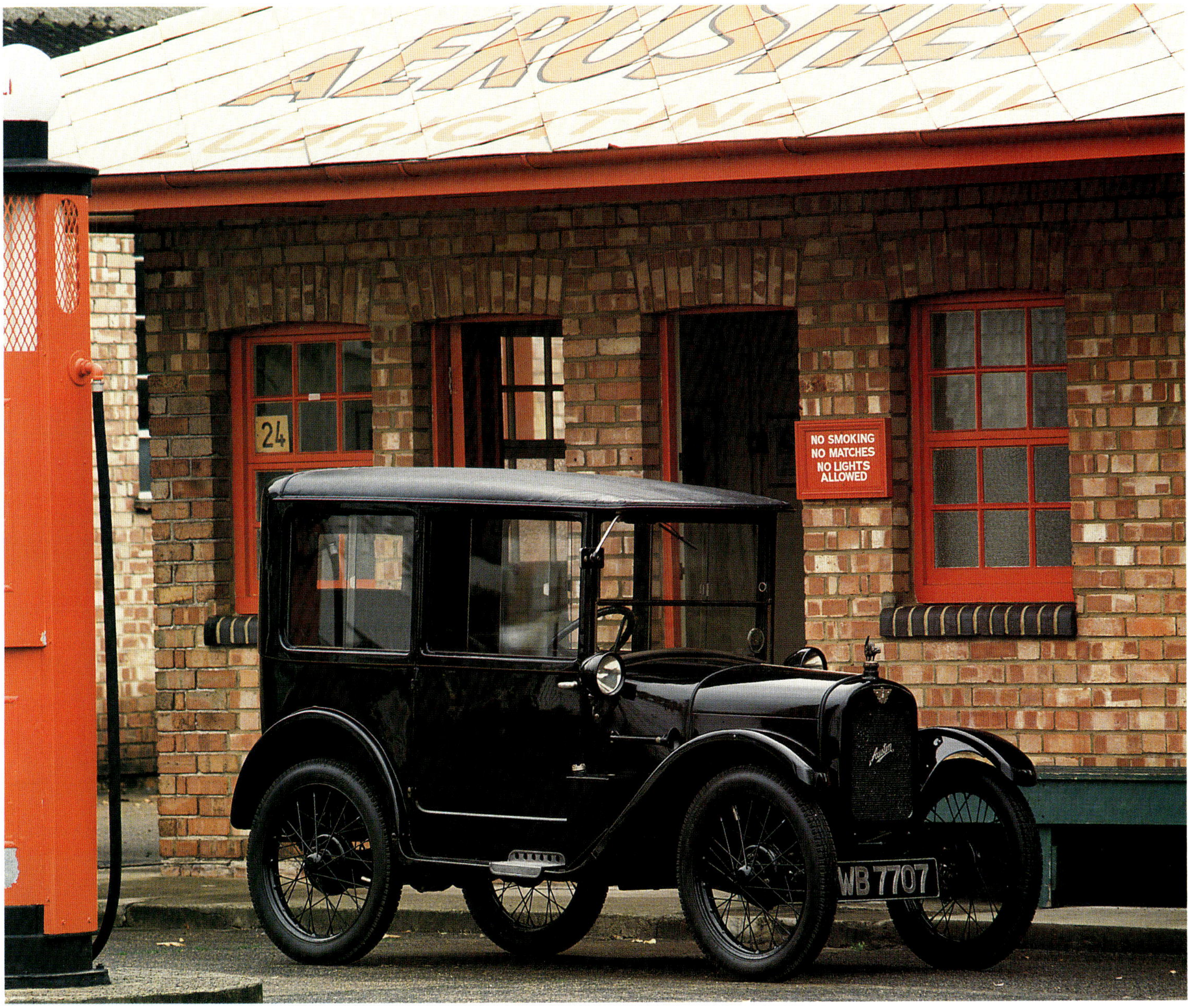
24
NO SMOKING
NO MATCHES
NO LIGHTS
ALLOWED
WB 7707

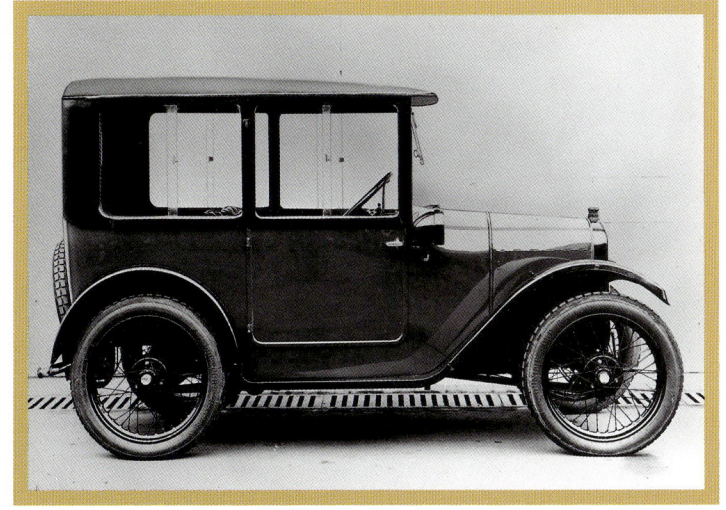

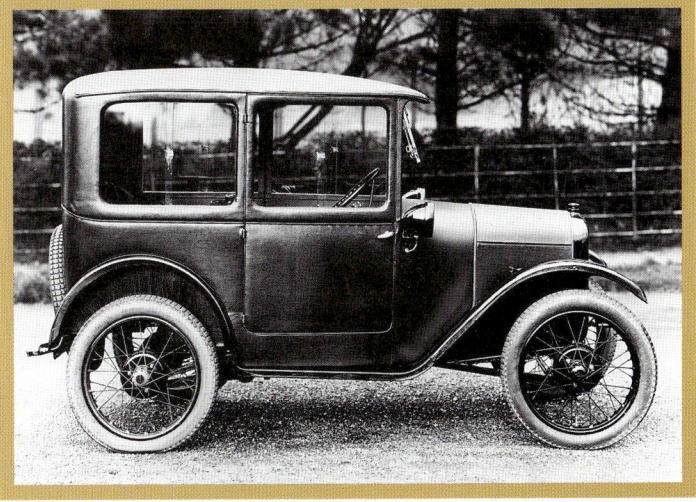

ORIGINAL AUSTIN SEVEN

The AUSTIN SEVEN "FABRIC" *Saloon* "*Easily the best small car in the world.*"

THIS model has a fabric body. Is light, silent and non-drumming. Single-piece windscreen opens at bottom to any angle. Very wide doors give easy and convenient access to all seats. One glass in door slides, the other lifts to any position, which gives the driver wide choice for vision, signalling or ventilation.

Superior upholstery and fittings. Dainty and inviting interior.

SPECIAL MODEL—WITH TRIPLEX GLASS.

Completely equipped, including electric starting and lighting, air strangler, electric horn, speedometer, automatic windscreen wiper, licence holder, shock absorbers, spare wheel and tyre, and blank number plates.

£150   *Special Model* £168
Complete at Works

PAGE 27

The fabric version of the RK saloon with black radiator, as seen in a sales brochure from 1928.

The same type of radiator shell used on this 1930 RK saloon (left and below) also appeared, with a stone guard, on later sports models. Wider doors, shaped around the wheel arch, were used for the RK saloon, and sliding windows were necessary due to this shape.

BODYWORK

This RK saloon (above) is a bit of an oddball as the nickel-plated radiator shell on a factory-bodied car normally means that the headlamps are mounted on the wings. It may have been made during a moment of indecision and probably dates from around August 1928.

Another variation on the RK theme (above) but rather more radical, with what is perhaps best described as a wrap-around sunroof. In all probability it dates from late 1929 or early 1930.

In March 1930 the RL saloon (right) was introduced. At first there were no scuttle ventilators but after a short while they began to be fitted.

A slightly later version of the RL saloon (below), seen in two-tone form in a sales brochure of 1931. By this time there are scuttle ventilators and louvred side panels in the bonnet.

ORIGINAL AUSTIN SEVEN

The fabric version of the RL, seen in a very attractive sales brochure, had its own designation of RG.

Another of Brian Davis' very original cars, this one an RP saloon first registered on 24 March 1934; chassis number is 189638 and engine number is 191665. Interestingly, both of his cars were registered in Breconshire and although there were nearly four years between them fewer than 1000 numbers had been issued in that time.

windows on later versions were a little larger. The third variation had a slightly longer body at the rear, thus increasing the leg room as the Seven gradually made its way to becoming a true four-seater. Pytchley sunroofs now had begun to be offered as an extra on closed Sevens.

RG In effect a fabric version of the RL, but with a necessarily tougher wooden frame; it followed a similar pattern of changes. It started with the plain bonnet and very short scuttle with ventilators squeezed in; louvred bonnets and long scuttles began appearing at the same time as on the RL, around chassis numbers 118750 and 128300 respectively. A sunroof or smoker's light could be fitted upon request. At this juncture readers may be amused to share a completely pointless piece of knowledge in learning that much of the fabric used for the bodies of Sevens went under the trade name of ZAPON, and was made by the Ioca Rubber and Waterproofing Company of Glasgow.

RN The first saloon body on the long-wheelbase chassis, at first made in de luxe form with a sliding sunroof and then a cheaper version was added without a sunroof. Bodywork was of pressed steel with raised waistline moulding and a pair of vertical ribs at the rear to conceal the panel joins. At last the doors could be wider, without the wheel arch encroaching upon them, and the better accessibility

BODYWORK

Most saloons were sold in de luxe form with a sunroof. Lighting is exactly correct on this 1934 RP.

Owned by David Brooks, this is the cheaper version of the ARQ Ruby. Chassis number is 215879, engine number is 214973, and the car was first registered on 25 March 1935. The radiator slats are painted black on this model, but finished in silver on the de luxe car with sunroof and bumpers.

No bumpers and a fabric-covered roof on this 1934 ARQ Ruby saloon. The owner has covered every eventuality by retaining the original rear light while fitting more modern examples on the wings, as well as a pair of reflectors.

to the rear seat was enhanced with little footwells for the occupants' feet. Most of these bodies had a scuttle-mounted petrol tank, but in the late summer of 1932 the tank was moved to the rear.

RP Announced at the time of the Motor Show in October 1932, production began at about chassis number 162600. The twin seams at the rear of the previous saloon disappeared due to the factory finding a neater way to join the panels, and the body was fractionally narrower across the rear seat. This is the most common of the chromium radiator saloons with over 14,000 produced during 1932 alone – around 80% of these were the de luxe version.

ARQ The first version of the Ruby saloon, from

31

How a sales brochure of 1935 extolled various virtues of the Ruby's design.

FOR EASE OF CONTROL AND RIDING COMFORT

All the instruments and controls are in front of the driver and in the nearside of the fascia panel there is a roomy receptacle for small parcels, etc.

THE factors which contribute to the comfort and enjoyment of Austin Seven motoring are far too numerous to describe and illustrate in their entirety. Here it is only possible to feature a few of the outstanding luxuries incorporated. But these can be taken as typifying the advantages embodied in the Seven. They go to prove the forethought devoted to providing for almost every motoring requirement within the compact proportions of this famous car, from the simple maintenance operations to the manipulation of the controls. There are thus no regrets or shortcomings experienced on investing in an Austin Seven.

Each large rear quarter window opens at its trailing edge to provide draught-free ventilation. Triplex glass to front screen and side windows.

Each front seat is separately adjustable and the front passenger or driver can instantly adjust the position to achieve maximum comfort without alighting. The leg-room is particularly generous and the wide door allows easy access to front or rear seats.

The rear aspect of the saloon and cabriolet is particularly neat. The luggage carrier will carry 56 lbs. The accessible filler for the rear petrol tank has a quick detachable cap. The large tail-lamp adequately illuminates the number plate and incorporates a stop-light.

The sensibly elegant lines, the roominess, the excellent visibility afforded by the large windows and slender pillars, and the wide aperture provided by the Pytchley sliding roof when fully open, are all revealed in the view below of the Ruby Saloon.

The steering wheel centre incorporates the horn button and direction indicator trigger—the two most frequently used controls. The latter is automatically moved out of contact on the return rotation of the wheel after cornering, and the indicator then drops back into its slot within the door pillar.

The luggage carrier folds into the spare wheel compartment when not in use. With the cover replaced the clean and assured curves of the rear panel are uninterrupted, and the full single-bar bumper ensures adequate protection. There are no awkward corners to render washing and polishing difficult.

Page 8 Page 9

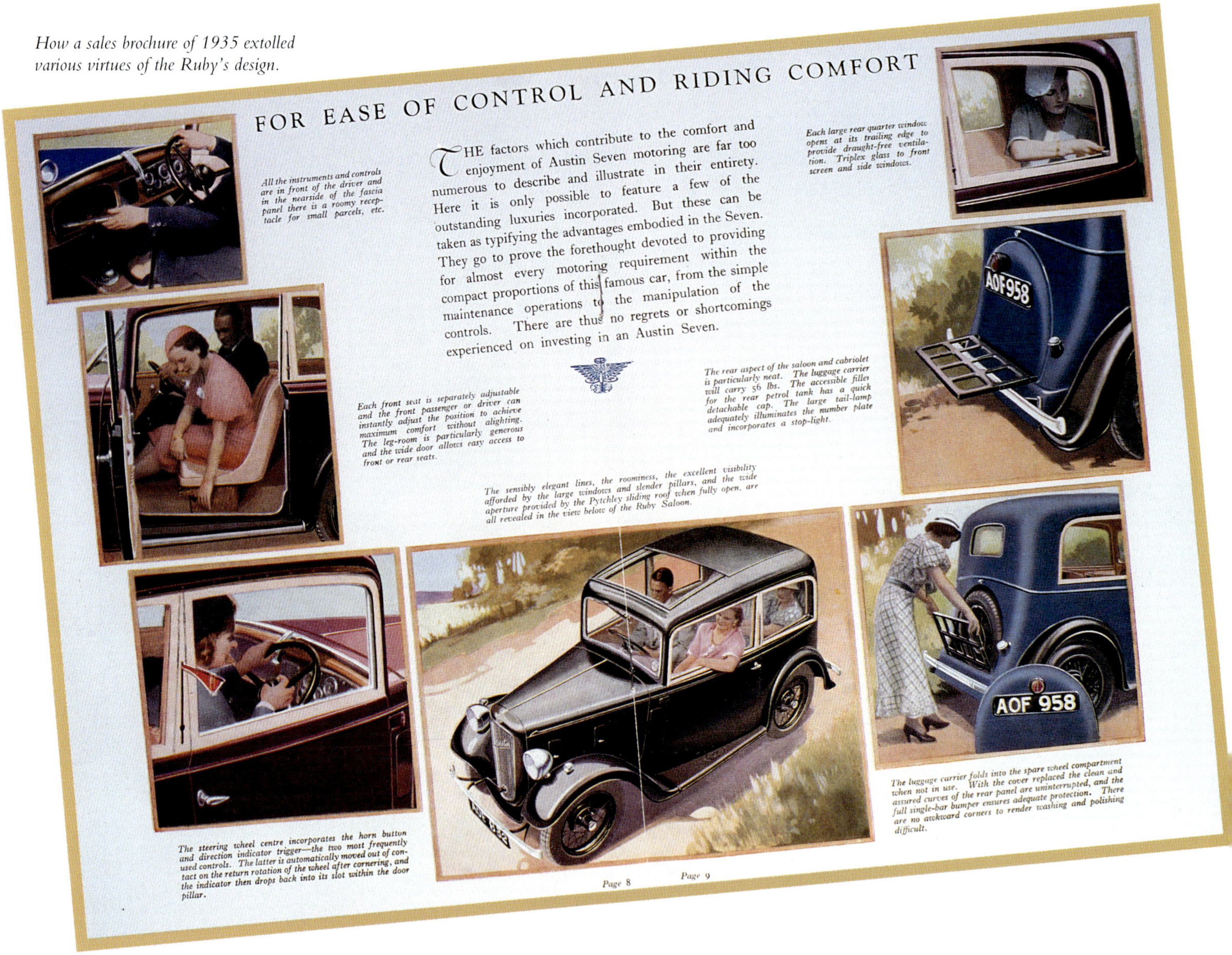

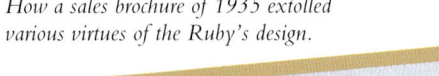

chassis number 198596 in July 1934. Built on the 'low-frame' chassis, it had a cowled radiator, a bonnet with hinged side vents, and a more curvaceous body with semi-swept tail incorporating spare wheel cover – the look of the Seven was completely transformed. The rear side windows were rather flimsy hinged affairs that suffer from rust; a Pytchley sliding sunroof was fitted as standard. A rarer version, which sold for £112 as against £120, was fitted with a smoker's light instead of a sunroof, and also had no bumpers.

ARR The New Ruby was introduced in midsummer 1936 at chassis number 249701. Chief distinguishing features from the previous model were the increased slope to the windscreen, radiused window surrounds with more substantial pillars, and winding rear side windows. Two models were made as before – de luxe with sunroof and the bumperless version with smoker's light. Prices rose that year to £125 and £118 respectively.

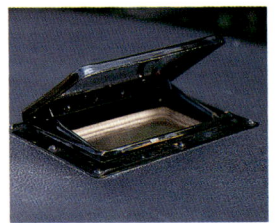

The roof ventilator or 'smoker's light' that was fitted instead of a sunroof on the cheaper Ruby.

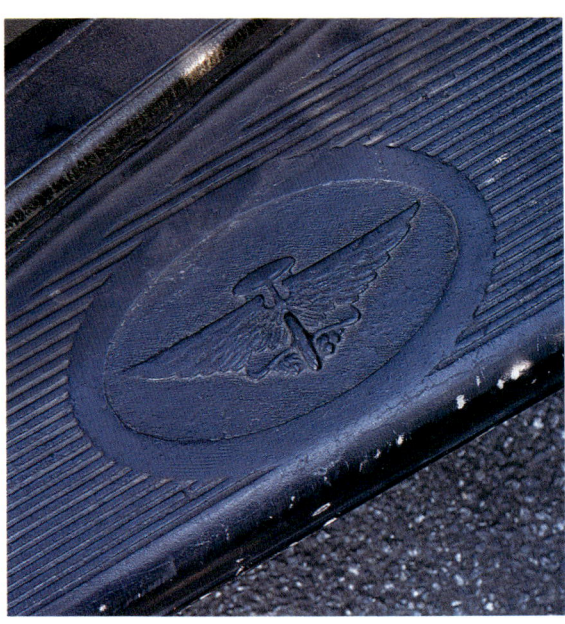

The moulded rubber running board covering carries the Austin motif, seen here on an ARR Ruby; replicas are made but the originals should be retained if possible.

BODYWORK

Dr James Anderson owns this ARR Ruby which was registered on 3 March 1937; chassis number is 266353 and engine number is 265663. From the front the main difference from the ARQ Ruby is the increased slope of the windscreen. In order to ease manufacture and clean up the design, body pressings around the side windows were radiused, with the flimsy hinged frames for the rear passengers discarded in favour of winding ones; the doors are also different. Originally only the central rear light would have been fitted.

ORIGINAL AUSTIN SEVEN

CABRIOLETS

AC Pearl A folding-head version of the ARQ Ruby with abbreviated side windows to allow space for the hood mechanism and external hood irons.

ACA New Pearl Made concurrently with the New Ruby, this was less distinctive than the AC Pearl. External hood irons were not fitted and larger side windows gave it almost the same profile as the parallel ARR saloon.

Bulky cabriolet hoods always presented a problem for the manufacturers when folded (above), and the AC Pearl is no exception. The fabric coverings, both inside and outside the upper window frames, are correct. The lamps on the rear wings are a later addition. The rear side window (left) was abbreviated to accommodate the hood mechanism; the leather strap ensures that the hood remains rolled back in this semi-open guise. A new hood to original specification (far left), with two-piece rear window frame sandwiching the glass, correct headlining and various straps.

The ACA Pearl cabriolet, introduced in September 1936, was less of a departure from the saloon than the AC version. Full-size rear side windows are immediately obvious.

BODYWORK

TWO-SEATERS

1929 Available for this year only, around 100 of these were made. The doors carried partially over the rear arches in the same way as on contemporary saloons; short bonnet with scuttle ventilators.

1930 Very similar, but now with a long bonnet and short scuttle; in 1931 the bonnet was louvred and scuttle vents were added. In common with the 1929 car, the top of the tail hinged at the front to expose the spare wheel and small luggage space.

PD Built on the long chassis, this lost the lithe looks of its predecessors, with rectangular doors and a less shapely rear upon which the spare wheel was mounted; any luggage was placed in the tail from within and the seats tilted for this purpose.

APD Opal In the late summer of 1934 the PD was renamed, carrying on with chromium radiator

The very rare 1929 two-seater (above), the first of the type, had a short bonnet and scuttle ventilators. The second type of factory two-seater (right), with a short scuttle and long bonnet – and here with the scuttle ventilators more typical of a 1931 model. This car, owned for many years by Michael Norris-Hill, was registered on 16 September 1930 and has chassis number 116101 with car number B1 8974.

Period view of 1930 two-seater (below) shows how the hood looks a bit of an afterthought on these cars; note the absence of scuttle ventilators. Charming 1931 brochure image (below right) shows added bonnet louvres and scuttle ventilators for two-seater.

35

This 1934 PD two-seater belonging to Trevor Turpin has chassis number 188191 and was registered in January 1934. This was the cheapest Seven which at one point was sold for the magic £100.

Brian Sutton's APE Opal has chassis number 282930 and engine number 284134, and was registered on 1 March 1938. It is a late car with large wheel centres and trafficators inset in the body behind the doors. Access to the tail space is only from behind the seats. A 'de luxe' pre-war number plate is flanked by duplicated rear lights to satisfy the MoT. The car looks rather more handsome with the hood down.

BODYWORK

An early version (top) of the Mulliner-built Military Seven, with short bonnet and scuttle-mounted headlamps.

Military Seven's boxy tail required no styling genius, but provided plenty of storage space for kit.

MILITARY

1929 There were several variants of the Military Seven, while the army additionally tried out modifications – including Horstman suspension and other devices – for different purposes. Mulliner of Birmingham built the tourer bodies with only one door, on the passenger side. The hind quarters were a good copy of a council house coal bin, but must have provided the 'squaddies' with plenty of storage space for their kit. Around 150 were made over two years.

1932 The bodywork on this version, also made by Mulliner, had different dimensions due to it being built on the longer chassis, plus the contemporary louvred bonnet and short scuttle, but otherwise it hardly differed from the previous batch. Once again specification varied depending upon the intended field of operations, which for the British Army ranged from Salisbury Plain to the Khyber Pass. Sales literature offered the little beast with left-hand drive, but it is doubtful that any despot was sufficiently impressed to equip his forces with a herd of them.

Austin itself produced a number of tourers for the Royal Corps of Signals at around the same time. This had a normal louvred bonnet aft of which the scuttle was widened by some 6in on either side, without any attempt to produce a gradual increase; this was necessary in order to mount the bulky radio set in front of the passenger. The car had two doors with the spare wheel mounted on the flat rear panel.

The two-seater evolved from PD form into the APD Opal midway through 1934, but with minimal change. This sales brochure comparison with the open four-seater shows how the APD Opal continued with a chromium radiator and louvred bonnet after other Sevens had adopted a new look.

and louvred bonnet at the time when other Sevens had been given a completely new look. Was this to use up remaining parts perhaps? The same car was made in military form for use in India: these had differing specifications from contract to contract, such as stronger suspension, 17in or 18in wheels with larger section tyres, and better cooling.

APE Opal In July 1936, from chassis number 247765, the two-seater was given the cowled radiator and bonnet with flap vents. At first there were no trafficators but midway through production these began to be fitted on the scuttle. In the summer of 1937 wheels with larger centres were introduced on the passenger car range and by now the trafficators were built in to the body behind the doors.

The Sports of 1924 had a slightly Gallic influence about the wings; the problem of where to put the rear number plate was solved by placing it on brackets behind the rear axle.

The same 1924 Sports with its lofty hood raised. Amusingly, the object on the tail in this publicity shot appears to be a wallet, left there by a distracted photographer?

SPORTS

50mph Sports The first sporting Seven was a very pretty little car which had a long, tapering tail with a rounded end; slightly cut-back doors (with a sloping rear edge) and rather skimpier wings than the tourer completed the effect. Produced until 1926, with some 300 believed to have been built.

E Super Sports This pointed-tail two-seater with no doors and cut-away sides was the forerunner of the Ulster, but without that model's lowered suspension. The spare wheel was housed crossways in the rear, reached by removing a cover. Production has been put at only five during 1927/28.

EA Sports The legendary Ulster. Similar pointed tail to the earlier Super Sports, but the top has a gentler downward curve which results in the vertical end being deeper; the spare wheel is mounted in the same fashion but the cover is smaller. Long louvred bonnet and no doors: the body has cut-away sides almost forming a shallow V as opposed to the U-shaped cut-outs on the Super Sports. There is a shallow fold-flat windscreen and a rather impractical hood. The car sits lower than previous Sevens due to the special front axle and lowered suspension. Standard tourer wings were employed, but slightly modified to suit. Around 300 were produced.

This heavily retouched factory picture of a 1927/28 Super Sports shows the different profile from the Sports (Ulster); these cars sat higher than an Ulster as they did not have a dropped front axle.

BODYWORK

The deteriorating banking of the only complete section of track remaining at Brooklands provides a backdrop for this 1930 EA Sports, or Ulster. Cars did not leave the factory with this binding around the exhaust pipe. Suitable for track or road use, the Ulster was supplied with full road equipment. Cars did not leave the factory with louvred bonnets, but this one was modified very early in its life. Headlamps, as can be seen clearly here, did not bolt to the cross-bar as might be expected.

The EB 65 belonging to Chris Keevil, first registered on 17 April 1934; chassis number is 193400 and engine number is 194013. The doors tend to distort on these aluminium-bodied cars more easily than on the later steel-bodied Nippy. Trafficators were not fitted to this model and the ribbed headlamp glasses are correct. At the bottom of the stoneguard is a Junior Car Club badge – just the sort of club, centred at Brooklands, that a sports Austin owner might have belonged to.

EB 65 After the raceworthy Ulster, the factory made no attempt to continue along these lines and the 65, although pretty and workmanlike, was a pale shadow of its predecessor. The rounded tail had an opening boot which only housed the spare wheel and tools; the separate luggage compartment was reached from inside the car by tipping the seats forward. In common with all sports Sevens up to this time, the body panels were of aluminium, but fully-valanced steel cycle wings were now introduced on sports models and these continued to be used on all further varieties. Some 275 65s were made.

AEB Nippy Almost identical to the 65, but the body panels were now of steel. Both 65 and Nippy enjoyed some popularity in Germany and were available there in left-hand drive form, as was the Speedy. The Nippy remained in production until 1937 and some 800 were produced.

Not quite as sporting as it may appear, but the AEB Nippy was nevertheless a very pleasant little sports tourer. Owned by John Clark, this car was registered on 24 March 1936; chassis number is 239669 and engine number is 240716.

BODYWORK

Althogh the boot of the Nippy looks large, it only holds the spare wheel and tools; any luggage is loaded into the forward compartment by tilting the seats. The rear lights are not strictly original but have been tastefully arranged to comply with modern regulations; sidelights should be chromed.

The AEK Speedy was in the idiom of many other small sports cars of the early 1930s, trading on the Brooklands image – all show and not too much go – but nevertheless a very desirable piece of property.

41

Original Austin Seven

In common with other Sevens, even the sports models had begun to put on weight in the 1930s and the wheels were now looking quite spindly in relation to the rest of the car. It could be argued that if the tail of the Speedy were less droopy, it would give the whole car a lighter appearance.

A stop to refuel a Speedy might prove to be an awkward experience while running with the hood up. All the best 'racers' had bonnet straps, and this is how they were arranged on the Speedy. There are experts who believe that some Speedy models had curved glass at the top of the windscreen panels, but this version is more normal; the single wiper is driven from a centrally mounted motor.

EK 75 Looking as if it was intended to be a successor to the Ulster with its aluminium pointed-tail body and impressive racing filler cap, the Speedy, as the factory must have realised, was in reality to have the role of boy racer rather than racer. The days of achievable success in racing by a small sports car producing 23bhp were gone, although a modified example did run at Le Mans.

AEK Speedy The same car as the 75, which was renamed to fall in line with the factory's new identification system. There has always been uncertainty regarding the number of 75/Speedy cars made, but a good guess would probably be 60 cars. It was discontinued due to lack of sales in 1935, but in the contrary way of things is now one of the most sought-after models.

BODYWORK

The rare and very desirable Type B coupé owned by Shotaro Kobayashi; chassis number is 97502 and the car was registered on 23 October 1929. This was really a two-seater but there was a small third seat crosswise in the rear. The rear of the Type B coupé is completely different, but the rest of the bodywork, including the doors, looks as though it was from an RK saloon – but panels are not interchangeable.

COUPÉS

1923 An extremely early tourer was converted to a coupé for Irene Waite, Austin's daughter. Up to the windscreen it was unchanged, but aft of here was a lofty two-seater body with sliding windows in the doors and coach handles. Disc wheels were fitted.

1924 Another one-off coupé or doctor's coupé, quite similar to the first, but the rear roof line was more angular and the headlamps were mounted on the front wings rather than the scuttle.

1925 Yet another coupé was tried, this one with an even shorter body and no rear side windows. It was almost pure tourer to the rear of the doors and had, in common with the previous designs, a small luggage boot.

ORIGINAL AUSTIN SEVEN

The short-lived Seven Taxi of 1925; the third hinge on the door may have been intended to resist wear and tear from heavy-handed cabbies.

Type B The only coupé to go into production was made between 1928-31. Aesthetically pleasing from almost any angle, it featured long doors which ended over the wheelarches. At first there was a short bonnet, but later cars had the long plain bonnet; the scuttle changed to match. The lower body was metal-panelled and the upper body fabric-covered. The two front seats were supplemented by a single crosswise rear seat. Although it was produced in fair numbers with just over 500 leaving the factory in three years, the coupé is now extremely rare.

Taxi In 1924 someone at Austin had the bright idea of producing diminutive taxis from the Seven and to this end several were made by the simple expedient of constructing a hard-top, with sliding rear side windows, on the contemporary tourer. Arrangements were made for these to work in London for an experimental period and for several months they plied the streets. However, with just one side-facing single seat at the rear and space for a hat box beside the driver, it was realised that more often than not those travelling by taxi either had more luggage or were not alone. The cars were returned to the works to be used for other purposes and the project was quietly dropped.

Commercial Traveller's Car Another good idea from the factory, this was built on the earlier AB tourer. A hard-top with two little rear doors was constructed over the back of the body while the driver was protected from the elements by a fabric top which could be furled. The interior was fitted out with shelves for samples and a container replaced the passenger seat. Although this vehicle appeared in contemporary motoring magazines and was advertised as well, for some reason the idea did not catch on and few were built.

VANS

AB, AC and AD From 1923 until 1927, Austin offered a delivery van that was built by Thomas Startin in Birmingham, using the contemporary tourer bodywork back to and including the doors. Rather more practical than the ill-starred Traveller's car, it had a pair of full-length rear doors containing oval windows, the shape of which was echoed in the profile where the panel sides swept around to the roof and doors – a most charming vehicle was thus produced. Weather protection was provided for the driver and mate by means of roll-down side curtains.

AE This later version of the Thomas Startin van was built using the 1929 tourer as a basis. It was very similar to the earlier versions, but lost some of the flowing curves that unified the previous design. The complete range of these vans sold well during the seven years until their production finished at the end of 1930 (with a few sold in early 1931), and somewhere in the region of 2000 were made of all types.

RK Austin itself built this van using the RK saloon as a starting point, but without this model's wide doors; awkwardly slim rectangular doors with sliding windows were provided to ensure the greatest possible space for goods. The spare wheel was mounted on the roof. Possibly because the inconvenience of the doors was too much for aspiring tradesmen to endure, it was a slow seller and only about 100 were produced.

RM With this van, based on the RL saloon, the factory learned its lesson regarding door size, by sacrificing some load space for ease of access via the wide doors used on the saloon. It must have been the right decision because van sales shot up to over 1000 in 1931.

BODYWORK

The first Seven van, the AB of 1923 (top), looked charming with its rounded side panels and oval windows in the rear doors. The final Seven van, the AVK based on the 'New Ruby' (above), appeared more utilitarian, but there was more load space.

RN The dilemma of the previous van of load space versus door size was alleviated with this model, which was based on the long-wheelbase RN saloon; rectangular doors and a longer body resulted in a carrying capacity of over 5cwt or nearly 50 cubic feet. Over 2000 were sold.

RP An updated version of the RN van, introduced from around chassis number 180000 in the early autumn of 1933, it incorporated the refinements of the RP saloon, such as its four-speed gearbox. As on previous commercials made by the factory, the van part of the body was panelled in aluminium on a wooden frame.

AVH This final update on the RN and RP vans appeared after the 'jewel' series of cars had gone into production. As it was a mere van, the factory obviously saw it as a convenient way to use up bits and pieces left over from the now redundant tourers and saloons. Considering the scarcity today of Seven vans, it is perhaps surprising that a total of 6000 of this model and its predecessor found customers.

AVJ and AVK These were based on the Ruby and to a certain extent followed the development of this series, although, as with other Seven vans, there were anomalies. The bodies were now panelled in steel over a wood frame with the roof covered in fabric. The spare wheel was carried under the floor and projected slightly through a cut-out in the rear doors, the latter each having a rectangular window with a pair of louvres above. Sales were very good, but on 3 March 1939 the last van and last Austin Seven, chassis number 291000, emerged at the end of the production line.

Note This information on vans is for guidance only: different specifications occurred when an order was large enough to make them worthwhile.

ORIGINAL AUSTIN SEVEN

INTERIOR TRIM

This early interior from a 1923 AB tourer is all original with the exception of the door check straps and carpets.

The driver's seat of this 1926 AD Tourer (right) has suffered a little but for an original interior to survive, even in this state, is remarkable. Compare this picture with that of the 1923 car to see the subtle differences in the front seats. Nothing fancy about vintage door trims (below), seen on the AD tourer. Panels were still fixed with tacks; screws and cup washers came later. The original check strap is another notable feature.

The upholstery on all Sevens for the first few years was normally black Rexine (also called American Cloth), although some 1926-27 tourers were trimmed in khaki-coloured Rexine. Rexine is currently unobtainable new, but we can hope some enterprising individual will soon decide that there is sufficient market to have it reproduced. In the UK there are poor substitutes, but slightly better ones are available in the USA.

The seat cushions were sprung with horsehair or fibre covering and the pleated squabs had kapok

INTERIOR TRIM

Only the carpets are unoriginal but they give just the right feel to this beautiful 1926 R saloon interior. Rear seat squabs are higher on the saloon and the pleats seem to have a little more bulk; even the threshold strengthener in the back is fully trimmed. For one of the cheapest cars you could buy there were certainly some nice touches: the blind is operated at the rear rather than by a cord terminating above the driver on the offside.

wadding infill. Even though this material is more expensive and the covering is harder to obtain, it is important to avoid foam rubber or plastic during a retrim. Hood and sidescreens were also made from black Rexine (or American Cloth), not double duck as is often thought.

When Austin started to make its own saloons it all became considerably more complicated: Bedford Cord, Moquette, Rexine and leather were all used. The best advice, in the absence of even the smallest vestiges of original materials remaining in your own

Original Austin Seven

There is a new headlining, but this 1930 RK saloon has the correct panels surrounding the rear window to harmonise with the upholstery panels around the side windows.

Soiled it may be, but this is exactly how a 1930 RK saloon door trim (bottom left) was made 65 years ago. Was this the inspiration for the same system of wire pull for the door latch on the early Mini? The perfectionist might like to try to find some of the material, seen in close-up, or even go to the expense of having it woven! Door panel close-up from 1929 RK fabric saloon shows that Rexine trim was also used on saloons. The reinforced flap is for pulling the door closed.

INTERIOR TRIM

Numerous details visible on this 1930 AE tourer include the floor-mounted starter switch, leather-bound slot for the handbrake lever, clamping screws for the sidescreen pegs in the doors, the construction of the control levers in the centre of the steering wheel, and the perforated steel floor pan side on the scuttle.

Rear seat changed little on the tourer – this is a 1930 AE version – during the early years. Rear sidescreen socket is painted black to match Rexine trim, wheelarch trim is not merely wrapped around but has piping (giving it a sharp delineation at the join), and floor pan is still of riveted construction. The hood frame, also seen on the 1930 AE tourer, was strapped down onto a bracket at either side when folded.

car, is first to try to seek out a car of the same type and year with original interior and, second, to contact the British Motor Industry Heritage Trust at Gaydon in Warwickshire. Although the records are sadly incomplete, this archive has the Austin Seven build ledgers for the years 1929-31 and for a search fee should be able to tell you the types of trim and corresponding colours that were used during those years – even if your actual car is not listed. For other years there is always the 'get-out' that the factory thoughtfully provided at the time by stating in much of its sales material that 'any reasonable colour of paint or upholstery could be supplied at an extra charge' – but it would be very wise to exercise self-control if the latter course is taken.

By the time the RN saloon appeared in 1931,

Beautifully retrimmed 1934 RP saloon, owned by Nick Turley: a good attempt has been made to replicate the flecked carpet, but the red is too bright.

Original seating from a 1930 AF tourer. Rear seat cushion, before the advent of Moseley 'Float-on-Air' seats for the Seven, shows arrangement of springs, hessian and Rexine. Front seats with original trim (both were to be preserved), one partially stripped to reveal steel pressing and construction, and filling of pleats.

INTERIOR TRIM

Rear side window (far right), on 1934 RP saloon, still has a separate panel surround. The mat covering the transmission tunnel is incorrect, and if Moseley 'Float-on-Air' cushions were fitted in the seats they would look a little less board-like; but shape of the seat squabs is excellent.

Interesting details from an unrestored 1934 RP saloon. Apart from the check strap, an absolutely original door trim (below): dark blue carpet flecked with black stops short of the door bottom; door latch pull is now chain covered with leather to replace weaker spiral-wound wire used earlier. Close-up of an original door check strap (right, above) shows a detail that is often hard to reproduce. Tool box cover under passenger seat (right, below) shows how finish was sparing in some places. The various textures of an original floor mat (bottom right), and carpet used only on the side facings.

51

Original Austin Seven

The interior of the Speedy (above), this one a 1934 example, has the look of a thoroughbred racer, with only the steering wheel boss giving the game away to the casual observer. By the mid-1930s the rear seat (left) of the tourer – this is a 1935 AAL Open Road Tourer – had become a more sumptuous affair. The indentation in the hood rail is to provide clearance for the wiper motor. Rear compartment of 1934 ARQ Ruby fixed-head (below), showing rear blind running on cords on either side.

INTERIOR TRIM

Battered but unbowed – the gorgeous original interior of a 1936 AC Pearl. The carpet was no longer flecked, but the same mixture of rubber matting and carpet continued from previous saloons. The draught-excluding rubber beading around the door apertures is Rexine-covered. The small catch for opening the rear window can just be seen. On the door, radiused marks on the carpet are from 60 years of compression against the body; the elasticated pocket has become a little tired.

Original Austin Seven

Although this is another Pearl, an ACA from 1937, it is the same as the contemporary Ruby. Interior window surrounds have a more utilitarian look than on previous models. Splits on the celluloid covering of the steering wheel boss are quite common. Rear compartment view shows that hood mechanism now lived inside and was simplified. Door and side panels were of Rexine, with leather for the pocket and the covering of the door latch chain. Note that the chromed pull handle has changed in style from 1936 to 1937.

Sun visor contrast. The rather lethal-looking tinted visor fitted to the AC Pearl, and the Rexine-covered version used on saloons, seen here on a 1937 ARR Ruby. Moseley 'Float-in-Air' cushion from inside the rear seat of a 1937 ARR Ruby.

54

INTERIOR TRIM

ARR Rubies and ACA Pearls had this rather awkward-looking handle (right) to open the rear side windows. These trim panels are all original but have just faded differently. Brown Bakelite covers for the door catches are fairly easily broken and originals are becoming scare. Seen on a 1937 ARR Ruby (far right) are the correct handbrake gaiter, gearbox cover, rubber grommet and matting; the flap retained by poppers is for the ventilator.

New door trim and original centre section of 1938 APE Opal. Door trims on the Opal and Tourer are somewhat bulky as they are used to store the sidescreens. Rear section of the sidescreen hinges forward for ventilation and hand signals. The door trim flap is secured by the two poppers at the base and hinges up to expose the sidescreens; the door pocket is stitched onto this.

things had calmed down a bit and largely these cars were trimmed in leather to match the body colour, although the AH tourer broke new ground for open cars by offering dark blue Rexine trim to supplement black. Headlinings on saloons were almost always beige but there were undoubtedly variations. Moseley 'float on air' seat cushions were used from 1928; most survivors have problems holding their wind, but these cushions are now being reproduced.

For the economy-minded, a passable substitute can be concocted from motorcycle inner tubes.

Rubys and Pearls normally had leather trim to match the bodywork, with some Rexine panels; the headlining on the Ruby was beige, or blue-grey on blue cars. Open Road Tourers and Opals had Rexine trim, whereas Sports Sevens were trimmed in leather apart from the Ulster, which had Rexine seats unless otherwise specified.

INSTRUMENTS

The first Sevens had no instruments whatsoever apart from an ammeter integral with the lighting switchboard. Not until the spring of 1924 was a speedometer offered as an extra and only the best part of a year later, in February 1925 from car number A1-8250, did the speedometer become standard equipment.

A flange-fitting silver-faced Smiths speedometer reading to 60mph was used, although it was possible to order the slightly more expensive version, with a trip in addition to total miles recorded, as fitted to the Austin Twelve Four. The speedometer drive was taken from a split pulley mounted on the propeller shaft via a spring belt to another pulley with a cable drive take-off. The secondary pulley shaft is carried on a bracket bolted to the chassis; the first type uses two bolts and a crimped end which clips over the frame, while the second has three bolts. The former has a grease cup and the latter a nipple.

In the autumn of 1927 this rather makeshift system was abandoned and a positive speedometer drive taken from the gearbox. The drive cable is shorter and rises visibly to the dashboard, but beware – there are different lengths depending on the speedometer's exact position. From the same date a Smiths black-faced, flush, 60mph, type PA speedometer was generally fitted, although it was still possible to specify the Twelve Four variety with trip. The customer could even have the clock and oil pressure gauge from the same source if desired, but this was unusual and any of these things would have incurred extra cost.

Sevens fitted with bodywork by outside manufac-

On the 1923 AB tourer all the factory fittings are grouped on the right-hand side of the dashboard, but here the CAV panel is partially obscured by the steering wheel. To its left is the oil button, to its right the horn push, and at the extreme lower right the magneto switch. A speedometer was not available from the factory when this car was made, but an accessory drive, by belt rather than spring, from the propeller shaft was available and was fitted to this car when new.

After the first production series, the dashboard was rearranged as on this 1924 AC tourer, with the CAV switch panel more or less in the centre; the oil button is directly in front of the driver above the patent plate, and the magneto switch remains low down on the right. An accessory dashboard light is fitted to the right of the switch panel.

INSTRUMENTS

On later tourers – this is a 1926 AD version – the windscreen frame was bowed to match the top of the scuttle. This is the optional Smiths speedometer with trip facility. The Jaeger clock would have been original equipment on rather more expensive English cars of the period, such as Humbers, but there was nothing to stop the owner of a humble Seven equipping his dashboard with one.

Dashboard of this 1926 R saloon has all the correct equipment: clock, dash light and ashtray are all admissible perid accessories. A CAV switch panel was used on early cars in this series, but by the time this one left the factory a Lucas panel had been introduced. Some owners liked a back-up to the electric horn and to this end a 'through-scuttle' horn, the bulb of which is seen here, was available from accessory manufacturers. The dished steering wheel is the rather more unusual one with spiral grips.

The black crackle finish seen on this 1930 AE tourer was used on both tourers and saloons of this period; seen here is the late type of oil pressure button and the non-dished type of steering wheel. This variety of switch panel/ammeter with half-charge and ignition warning light was fitted from the introduction of coil ignition. On two-seaters the dash differed in having a mildly radiused cut-out from the area of the choke ring to the patent plate, and the speedometer and switch panel were set a little higher.

57

Original Austin Seven

The crowded dashboard of the supercharged Ulster, or EA Sports. From left, air pump for fuel tank, boost gauge, fuel tank air pressure gauge, oil pressure gauge, 100mph speedometer, Lucas switch panel and ammeter, St Christopher medallion, water temperature gauge and rev counter. Spoiled trousers or skirts became a thing of the past from May 1930 when the leaky old oil pressure button was replaced by an oil pressure gauge, seen here on a 1930 RK saloon (left). A pair of miniscule cubby holes, one at each end of the dashboard, were now incorporated. Saloons have an extra, separate, moulding on top of the dashboard, making it deeper.

INSTRUMENTS

This dashboard, on a 1934 PD two-seater, was peculiar to two-seaters and tourers from October 1932 until the arrival of a cowled radiator on each model. The Bakelite steering wheel boss and barrel-type speedometer were introduced across the Austin Seven range at this time. Choke and starter pulls are below the dash, respectively below the patent plate and to the right of the steering column.

This was the equivalent saloon type of dashboard introduced with the RP in October 1932, with a panel carrying all the instruments matching a single large glove compartment.

The 65 (1933 car shown) and Nippy models have the same instrument panel as the saloon, but it is mounted in the centre of a wooden dashboard and flanked by a pair of cubby holes. The four-spoke steering wheel is a Bluemels; the three-spoke Frank Ashby variety was not used until around chassis number 200000. The windscreen opens from the top and the vacuum wiper fits the frame more neatly than the later electric type.

59

Original Austin Seven

It must have been reassuring for the Speedy owner as he crouched over the steering wheel, nearing terminal velocity, to see that he was not about to over-rev in top gear by means of this special combined rev counter and speedometer. Note the mounting of the wiper motor on the windscreen pillar.

The Ruby has a wood-effect painted dashboard and door cappings, well reproduced on this ARQ model. The trafficator switch is now in the steering wheel boss.

Instruments

The Open Road Tourer (this is a 1935 AAL) has the dashboard painted body colour, as well as a windscreen that opens further and is supported by telescopic struts. The flat rectangular section at top centre is where the windscreen opening handle would be mounted on a Ruby or Pearl. Close-up of the removable instrument panel – also shared with the Ruby and Pearl – clearly shows the barrel-type speedometer used until January 1936, when a conventional needle type was substituted.

turers were a law unto themselves and it would be folly to try to suggest what was original dashboard equipment on these cars. In many cases it would have been exactly the same set-up as the factory versions, but if a car has a more comprehensive layout then the instruments used should naturally be of the correct period.

From the start of production Sevens were fitted with an oil pressure button mounted on the dashboard in front of the driver. As long as there was a little oil pressure in the system once the engine was running, the button protruded from the bezel; at the same time it was prone to depositing the odd drip of oil on the driver's legs. This oil button was altered in September 1928; after chassis number 69104 it had a large flange and was secured to the dashboard with two screws.

Like the Ruby, the Pearl dashboard has a wood-effect paint finish, which looks strikingly dark in original form on this 1936 AC model. The large chromium-plated T-handle is for opening and closing the windscreen.

The oil pressure gauge that was a direct replacement for this oil button was manufactured as an accessory by David Harcourt, but early examples do not always have the maker's name on the face – you may find them marked only 'Linkula'. The type of gauges made by this company at this time, calibrated to 20psi, have a flange fitting with an enlarged upper segment and two screw holes. Their popularity led the factory to contract Harcourt to supply a flush-fitting version, fitted as standard from around chassis number 110900, during May 1930.

During September 1932 the Seven was given a general facelift and a pressed steel instrument panel was now fitted to the dashboard. As well as an ammeter, petrol gauge and oil pressure gauge, there was a black-faced Smiths Magmo drum-type speedometer of smaller diameter than previously, reading to 80mph. The oil pressure gauge, although still made by David Harcourt, no longer carried the word 'Linkula' and was calibrated only to 10psi.

The instrument panel on open cars was of a different shape to the saloon version and was mounted in the centre of the dashboard instead of in front of the driver. Dials and ancillaries were also arranged slightly differently on the two versions.

With the introduction of the Ruby the shape of the instrument panel pressing was slightly altered and the contents were repositioned; open cars now had their instrument panel in front of the driver, as on the saloons. In January 1936 the drum speedometer was superseded by a conventional one, still by Smiths and still reading to 80mph or, on some export cars, the metric equivalent. In February 1937, from chassis number 264705, an oil pressure gauge with higher calibration was introduced.

Sports Sevens very often had the same instrumentation as the saloons and tourers but in some cases it did differ. For instance the Brooklands Super Sports had an AT rev counter, driven from the front of the camshaft, which was calibrated with the road speed in top gear.

The normally aspirated Ulster optimistically had a Smiths 100mph speedometer with trip facility and a matching Smiths rev counter, and an oil pressure gauge as well as the normal Lucas switch panel with ammeter. Supercharged cars had the same layout but with the addition of a pressure gauge next to the fuel tank's hand pump on the left-hand side of the dashboard. The 65 and Nippy had the same instrument panel as contemporary touring models.

ENGINE

Engine on 1923 AB tourer shows how at first there was no cooling fan. Oil filler cap, on the crankcase web behind the Scintilla magneto, has an aluminium cap retained by a spring clip and is really quite awkward to fill. Dynamo cut-out is the first type with aluminium cap, and the horn is a Benjamin.

Designed just after the First World War, the Seven engine was to some extent a scaled-down version of its larger Austin relations. It was typical of the period in that the lower half of the engine was an aluminium crankcase casting to which was bolted a cast iron cylinder block and head. The Seven remained faithful to this method of construction long after other Austins had gone over to a one-piece cast iron block. Apart from taxis, even the long-lived Heavy 12/4's swansong was in the 1935 Hertford saloon.

In the late 1930s, aluminium crankcases with iron blocks were the province of manufacturers such as Bentley, Meadows and Alvis, so the reason why the price-conscious Sir Herbert continued to manufacture the engine of his baby this way is probably down to weighing the cost of retooling, pattern-making and possible teething troubles against a proven design that was economically viable. The Austin Seven engine, although lacking overhead valves, which by the end of its life were almost *de rigueur* for engines of small capacity, is nevertheless a very efficient unit constructed from materials of excellent quality.

As originally drawn up and manufactured, the bore and stroke were 54mm × 76mm, giving a cubic capacity of 696cc, but by the time the engine went into production, after car number A1-101, the bore had been increased to 56mm to give a capacity of 747cc, at which it was to remain for the rest of its life.

There is a one-piece cast aluminium crankcase and flywheel housing with an almost full-length horizontal web on either side that incorporates the engine mountings, and the web on the offside also includes the magneto platform and oil filler. The two-bearing crankshaft has big end journals and main bearings of 1⅙in diameter. Until 1929, the crankshaft in fact ran in three bearings as, in addition to the roller bearing at each end, there was a ball bearing next to the front one.

The camshaft, high up on the nearside of the crankcase, runs in a roller race in the centre with plain bearings at the front and rear. At first the thrust was taken by a button on the timing cover, but during 1924 a redesigned camshaft was introduced with a shoulder to take the thrust and an altered timing cover using a bush with oilways. On an engine with the first type of camshaft, the thrust is adjustable by shimming the button under the hexagon on the outside of the timing cover.

The timing gear case comprises a pair of aluminium

Provision was made for priming cocks but on this 1923 AB tourer engine domed brass blanks are fitted as they were not really needed – they were a hangover from earlier days. As there is no electric starter, the choke is operated from the front of the car by a wire passing through the radiator; there is also no windscreen wiper, so the inlet manifold is not drilled for a vacuum pipe. These early engines had the Austin script on the nearside crankcase web.

castings, inner (bolted to a machined section on the front of the crankcase) and cover, which as well as containing the camshaft drive gear has a spur gear, running on ball bearings, to drive the magneto. The dynamo is driven by a skew gear directly from the camshaft gear and sits in its own separate aluminium housing bolted to the top of the timing case, at right angles to the motor.

Attached to the top of the crankcase, by eight studs, is the cast iron cylinder block, whose valves and porting are on the nearside. The cast iron cylinder head is secured by 14 studs and sealed by a copper asbestos gasket. The first batch of cylinder head castings had an integral water manifold, but by the time the car was on sale to the public the small aluminium manifold retained by a set screw had been standardised. Until 1927 cylinder heads had provision for priming taps, the holes for which usually have brass plugs screwed into them. Cars other than very early models without a choke were sent out from the factory with these and owners could fit primers if they wished.

Another feature of very early cars was a cast aluminium sump and in conjunction with this a drip tray was sandwiched between crankcase and sump into which were let three circular brass gauze baskets. These were replaced by a pressed steel sump and simple gauze filter with reinforcing frame, also sandwiched between crankcase and sump, by the end of 1923.

For cars with no starter, a pair of cast aluminium semi-circular covers were bolted over the top of the flywheel housing, one half on the crankcase and the other on the gearbox; it must always have been the

intention to fit a starter. In the autumn of 1923 a delightfully simple, spring-return, pull-cord starting device became available to attach in place of these covers. It was not, however, a very delightful piece of equipment to use and, although the factory suggested using the starting handle from cold, one can picture all manner of odd contortions and oaths as the occupants wrestled with the thing if the little car did not commence first time. A major degree of sophistication was brought about at the end of 1923 when the factory decided that it might be best to spend some of its profit margin on an electric starter.

The starter motor, gear wheel, pinion and switch were all mounted in an aluminium casting which

A factory photograph, dated as 1924, of a magneto engine with gearbox and ancillaries.

ENGINE

By the mid-1920s the under-bonnet view – here on a 1926 R saloon – was becoming a little more crowded with the addition of a cooling fan and its attendant drive. Now that a starter motor was fitted, the choke could be operated from the dashboard; the inlet manifold has a vacuum pipe for the windscreen wiper.

The other side of the 1926 R saloon engine. A separate cut-out and fuse box were now employed; the awkward oil filler is fitted with a proprietary accessory, by Eversure, to overcome the difficulty of filling; the magneto, a BTH in this case, was one of the options available.

was bolted over the flywheel in the same manner as the previous starter. This remained in use until 1929 without change other than Lucas electrics replacing CAV during 1926. The Lucas starter switch had two mounting holes rather than the three of the CAV model, so the starter housing should be drilled correspondingly.

Engine lubrication was taken care of by a vane pump, driven by a skew gear at the rear of the camshaft, supplying lubricant under pressure to the plain bearings in the camshaft. The steel H-section connecting rods have two-bolt end caps with white metal bearings. The gudgeon pins are clamped in the little ends and articulate in the pistons, which give a compression ratio of 4.9:1 and are of aluminium alloy with three rings (two compression and a scraper below the pin). In order to further limit excess oil finding its way to the bores and possibly to the combustion chambers, flanged circular tin oil baffles, slotted for the connecting rods, were fitted into machined recesses at the top of the crankcase. With the increase in capacity after the first few cars had been made, a fresh size of piston was needed. This remained unchanged until April 1925, when the skirts were slightly reduced in size from engine number 9187.

The side valves are controlled by coil springs, retained by split collets, and operated via screw-adjustable tappets. Until 1924 waisted tappets with a radiused foot were used, in conjunction with guides to accept the latter, then from around car number A1-3000 these were changed to a shape that was much simpler to machine and remained in use to the end of production.

Very few modifications were necessary to the engine for some years except for trifling matters as a result of customer demand rather than mechanical necessity – such as a slightly extended starting handle shaft to allow owners to fit a radiator muff, from chassis number 59275 in April 1928.

Doubtless influenced by the widespread use of coil ignition on mass-produced American cars and the fact that considerable savings on the price of bought-in components could be made, Sir Herbert elected in 1928 to change the Seven's ignition system to coil. A good deal of redesign work was undertaken on the engine, starting with the crankcase, which had the magneto platform deleted and the oil

ORIGINAL AUSTIN SEVEN

The factory dates this heavily retouched picture of a coil engine and ancillaries as 1930, but the type of starter assembly indicates a date before October 1929.

filler hole moved to a bulge in the casting, centre off-side, into which a filler neck was fitted. The timing case had no longer to hold the gears driving the magneto, so it was reduced in size and the cover had a nose piece incorporated to carry the further extended starting handle shaft. This did away with the separate mounting for the starting handle shaft which until that time had been bolted to the chassis. As it was no longer possible to mount a pulley on the end of the crankshaft, the fan was driven by a pulley attached to the end of the camshaft, which in turn meant a camshaft with an extended spigot and remachined gear wheel.

The distributor was positioned at the end of the dynamo and driven by it. The dynamo was now flange-mounted with three bolts onto a redesigned housing instead of the clamp fitting used with the previous dynamo. The parts making up the fan assembly were all different due to the fan spindle locating straight into the dynamo housing rather than on a separate bracket, and the fan rotating in an anti-clockwise direction due to being driven by way of an intermediate gear rather than the crankshaft as before. The changeover period from magneto to coil engines was haphazard, but it is generally acknowledged that by chassis number 69001 the transformation was complete.

In the first month of 1929 the pistons were altered to four rings, from chassis number 77801, and then by late summer returned to three but with the scraper now stepped. In October 1929, from chassis number 97489, a stiffer crankshaft with big end journals enlarged to 1$\frac{5}{16}$in was introduced along with complete connecting rod assemblies to suit. The large rear main bearing housing, necessary due to the procedure of threading the crankshaft through the rear of the crankcase during assembly, was simplified and the felt seal gave way to an Archimedes thread.

Just after this, during the same month and from chassis number 99001, the starter motor housing was replaced by one of cleaner design which did away with the Bendix gear, the pinion acting directly on the ring gear and the switch transferring to the car's floorboards.

A new model of dynamo fitted with a more substantial gear wheel came into use from chassis number 110940, in April 1930. Shortly after this during May, from chassis number 112479, the cylinder block casting was improved by repositioning the tapped holes that accepted the studs for the tappet cover. Instead of being drilled straight into the water jacket, they were now in small lugs incorporated into the casting so that the water jacket was not perforated. As the studs were now off-centre, the newer

The engine number is stamped just below the front of the tappet cover; normally this was of pressed steel but this engine is in a 1932 Swallow and has the aluminium version produced for that company. By this time the Austin factory was no longer fitting bronze carburettors.

ENGINE

Everything on this 1930 AE tourer remains as the factory made it with the exception of the later jubilee clips on the water hoses. The dynamo now has an extension to house the distributor and its skew drive.

type of tappet cover had repositioned holes. At the same time the seal was altered from cork sheet to a peripheral type, and the Austin name was incorporated into the pressing of the cover.

The starting handle was given a deeper location on its shaft, which was itself altered to peg drive, in March 1931 from chassis number 130860. During the early summer, at around engine number 138000, the valve guides had the shoulder deleted from their design and at the same time the cylinder block received minor improvements to the water jacket.

Some owners had been troubled with the dynamo housing coming loose, so during September the set-screw location was improved and at the same time, from engine number 141868, the front bearing for the camshaft was enlarged, which necessitated a slightly modified crankcase (part number BA 140 was superseded by BA 155).

A completely new crankcase was introduced in September 1931, from chassis number 159534, and allowed the starter to be mounted on the offside of the engine. This did away with the untidy starter motor assembly on top of the flywheel, and a cover similar to that of the first cars, except that it was in one piece, now covered the top of the flywheel. At the same time the conical filter gauze in the oil filler neck was dispensed with.

On just over 100 cars, between engine numbers 159463 and 161385, the factory experimented with a piston having all the rings plus a row of drain holes above the gudgeon pin, but then reverted to the previous type. The material from which the pistons were produced was altered to low-expansion alloy at engine number 170274 in February 1933.

The Seven engine had always suffered from too much oil in the bores, a feature of the lubrication system which threw oil everywhere and not just in the right places. In another bid to stop oil passing the

Original Austin Seven

This engine is in a 1935 AAL Open Road Tourer. New wiring looms to the original pattern are available and orange fibre clips are correct. Starter and dynamo casings are cadmium-plated; 'Dependability' plate was fitted from new and is also available as a reproduction.

The new three-bearing engine exposed, as seen in a 1937 sales brochure but introduced in June 1936.

ENGINE

pistons they were once again redesigned in April 1933, from engine number 174921: the pistons remained in low-expansion alloy, but again carried all three rings above the gudgeon pin, the lower two having oil drain holes. Just after this the factory began a system of grading pistons and bores after machining to achieve even better tolerances. If you come across an engine from this period that has never been reconditioned other than at the factory, it will have a small diamond enclosing a number stamped opposite each bore on the top face of the block and on the top of each piston. These signified the tolerance, within .0005in, so that the components could be matched during assembly. During these piston experiments, the rear camshaft bearing wall thickness was increased, with effect from engine number 168230.

From chassis number 179368, rubber mountings were introduced for the engine, although vans did not benefit from this feature until chassis number 190743, in 1934. Loosening of the mountings was experienced at first so the set screws and lock washer were superseded by split-pinned nuts and studs from chassis number 183595.

The front crankshaft bearings were altered to double-purpose ball races from engine number 190766 during February 1934 and then, during that summer, with the introduction of the sloping radiator cowl on most models from chassis number 198596, engines fitted into these chassis received a timing cover with a greatly extended nose piece and longer starting handle shaft. The crankcase casting underwent some minor changes at the same time, including the oil filler aperture now accepting a threaded rather than a push-fit tube.

In June 1936, at chassis number 249001, the three main bearing engine was introduced, necessitating a fresh crankcase and various associated parts. The centre main was of split-shell type and at the same time the big end journals were reduced in width from 1¼in to 1in. In addition the centre bearing of the camshaft was changed to a plain shell bearing. The cylinder head combustion chamber was reworked to give a compression ratio of 5.8:1 and the spark plug size was now 14mm.

Meanwhile, the factory's search for improved oil control in the bores continued and at various times modifications were made. For instance, at engine number 207700 the old design of piston with oil ring at the skirt was reverted to, then at engine number 232501 the rings were machined with a convex face, and yet another variety of piston was used from engine number 252915. Connecting rods which would accept shell bearings appeared in March 1937 from engine number 268659; the correct bolts with 'flats' on them must be used in conjunction with these rods. At the end of 1937, from engine number 283618, the valves reverted to split collet retention rather than the drilled stem and peg they had used since January 1935, from engine number 213111.

The Brooklands Super Sports used a modified engine. Although not strictly a production model, it was sanctioned by the factory and available from dealers, although actually produced by Gordon England (see page 116). The engine was at first based on standard castings and components, and became progressively more modified into higher states of tune. By 1924 the crankshaft was altered to accept a pressure oil feed system, and some racing engines had stronger connecting rods and a crankshaft with circular webs machined from the solid. Most engines had their porting cleaned up and enlarged; cars made from the middle of 1925 used tulip valves activated by a high-lift camshaft.

The Sports model, introduced in 1924, was said to have a specially tuned engine to enable each car to achieve 50mph. What is more likely, I believe, is that the engines which performed best were taken from the production line and used in these lighter cars.

From 1925 the factory employed superchargers for some of the racing Austin Sevens, even building its own Roots-type blower. This was mounted at the front of the engine on a special casting. When the five works racing Super Sports cars, the true forerunners of the Ulster, were made at the end of 1927, they were fitted with Cozette superchargers. The engines also had counterbalanced crankshafts machined from the solid, cast aluminium sumps and more radical valve gear.

At the beginning of 1930 the Ulster was put into production in either unsupercharged or supercharged

Rear petrol tank necessitated a petrol pump, which is low down on the crankcase underneath the manifolding. Rubber buffer on top of the tool box lid is to prevent metal to metal contact. This 1937 AAR Ruby has the original under-bonnet finish – part body colour and part black.

form as the Sports. The supercharged engine used a magneto-type crankcase, with altered mounting feet and, due to the crankshaft having a pressure oil feed, undrilled oil jet holes. From the timing chest an additional drive was provided for the Cozette No 4 supercharger mounted on the side of the crankcase underneath the exhaust manifold. Additional cooling was achieved by using a water pump, in conjunction with a different cylinder head with larger water passages. The pump was driven by the same pinion as the magneto and was positioned at the front of the timing cover, which was deeper than normal and incorporated part of the water pump in its casting.

The five works Super Sports cars of 1927 had been fitted with special 1⅛in circular web crankshafts but the Ulster initially had a 1⁵⁄₁₆in crank, increased to 1½in in 1931. The big end bearings were pressure fed via an external pipe from the pump to a union in the side of the starting handle nosepiece, and thence through a drilled starter dog into the crankshaft, which was, of course, now drilled with exits only at the big end journals. Although the leather sealing washer on the starting handle shaft is assisted in its task by a secondary spring in this engine, it often

Only the supercharged Ulster had a magneto by this time (1930 in the case of this car). The water manifold was fitted back to front, hence the long convoluted hose. The supercharger and carburettor on the blown Ulster are all but invisible underneath the manifold. The large tank is for petrol and the smaller one for supercharger lubricant.

ENGINE

The two-bearing sports engines fitted to the 65, Nippy, 75 and Speedy (pictured) had this special crankcase which allowed clearance for the larger big ends; this type of manifold and carburettor were also used. Other side of 1934 Speedy engine shows that oil filler was now better placed. Note armoured wiring.

objects and oil leaks are common from this source.

Standard compression ratio was 5:1 and at first both types of Ulster had eight-stud crankcases, but with the greater actual pressure due to the supercharger the later cars were reinforced with 10 studs to secure the cylinder block, even on unblown engines. A longer camshaft is used in this engine as it drives the rev counter by means of a skew gear from the front; it was necessary to position the rev counter drive thus on the supercharged cars in order to clear the supercharger assembly. The supercharger itself is driven from the camshaft pinion. The tappets on these engines were not adjustable; instead, valve clearances were adjusted by altering the hardened steel caps on the end of the valve stems. There are two sorts, dependent on the type of valve used, either a cup or mushroom, and they were ground to size; valve heads are of the tulip type.

The late Ulster engine had a repositioned oil pressure relief valve to enable a stuck valve or other malady to be rectified without resorting to removal of the engine from the chassis, as on other models.

Although the factory had produced aluminium sumps with a capacity of around a gallon for use on various racing Sevens, especially those prepared for

71

long-distance events, the production Ulster used a pressed steel sump of normal capacity fitted with a circular cover which could be unbolted to attend to the gauze oil filter within. It was probably possible for the persuasive owner to obtain a cast aluminium sump as the factory was not averse to private owners' competition activities and even offered a larger supercharger for the Ulster. Around this time, cast components specifically for sports Sevens began to be produced with the casting prefix 9C on them, although this does not automatically signify that they are for an Ulster as this practice was carried on through 65, Nippy and Speedy models, and even to the works Grasshopper cars.

The normally aspirated Ulster engine did not have magneto ignition and therefore was based on the coil ignition crankcase, but neither it nor the supercharged motor is interchangeable with standard castings. As on the blown cars, the mounting feet were realigned to make allowance for the lower suspension, and there were no oil jet holes due to the pressure feed to the big end bearings. In addition the camshaft bearings were larger to enable a higher lift camshaft to be fitted and withdrawn. The rev counter drive came from the side of the camshaft as there was no need for a larger timing cover to cope with the water pump and blower, nor did the cable have to avoid the latter. Valve clearance adjustment and oil pressure relief valve arrangements were as on supercharged cars, the latter an added blessing if, as was often the case with both engines, they were run on vegetable oil, which can become glutinous after a while.

The factory allowed greater time for the assembly of Ulster engines. Some components were machined over, the bottom end was balanced, and greater care was taken with the fitting – although Austin had already reached far higher standards than many of its competitors.

The next generation of sports engines started with the 65 in 1933. The Ulster had been a proper sports racing car available to the public, but the 65 had no such pretensions and therefore the backward step of abandoning pressure lubrication to the big end bearings was taken. The engine, however, did retain the 1½in crankshaft with balanced bottom end, and the crankcase, otherwise of standard dimensions, is recognisable by having four bulges just above the mounting webs to provide clearance for the big ends. Until engine number 180614 these crankcases were sand-cast, but after that they were produced by the die-casting process then used by the factory. The cylinder block and head on the majority of Ulsters had been cast in Chromidium, identified by the raised diamond trademark, and this practice was continued for sports Sevens from then on. A high-lift camshaft was fitted, along with the necessarily larger bearing; tappets were of standard adjustable type but tulip valves were fitted, with double springs. A cast aluminium sump was used with a capacity of nearly one gallon; this was finned whereas sumps produced before this date were plain. For this engine the factory claimed a power output of 21bhp at 4400rpm.

At the end of 1933 production of the 75 engine began. This was very similar to the 65 except that it had the Ulster's pressure lubrication system, but without that engine's more convenient oil pressure relief valve. Output became a claimed 23bhp at 4800rpm.

Only a very few 75 engines were produced by the sand-cast process before the factory changed to die-cast for the sports motors. Rubber mountings were being introduced at this time on other models so the sports cars followed suit. The 75 model was renamed Speedy almost as soon as it had appeared, and the engine was available to outside firms for various concoctions such as the Arrow 75.

The 65 had now become the Nippy but the engine remained virtually unchanged until, some time around chassis number 240000, the engine fitted to the by then discontinued Speedy, with pressure-fed big ends, began to be used in the Nippy. From around chassis number 255000, during the autumn of 1936, it became possible, for a lower cost, to buy a Nippy with the normal three-bearing engine found in the Ruby.

In December 1936 the factory came up with a pressure-fed sports engine for the Nippy. This did away with the previous method of sending oil from the pump to the crankshaft and it was now distributed through the crankshaft by way of the plain centre main bearing. Connecting rods with floating gudgeon pins, the ends fitted with aluminium mushrooms to protect the cylinder bores, had been introduced with the Ulster engine and their use continued right through to this three-bearing sports motor. Power was quoted as 23bhp at the rather more leisurely speed of 4400rpm.

Between the wars Austin produced marine versions of some of its engines, the Austin Seven version going under the name of Thetis. These are instantly recognisable by the cast iron sump, cast iron housing for the starting handle and chain, plus the manifolding and pipework necessary on a marine engine. Very often the crankcase was also of cast iron. These engines continued to be built in stationary and marine form after the demise of the car, and saw war service powering everything from fire pumps to the Mk2 version of the Uffa Fox-designed airborne lifeboat. These latter engines are stamped AM (Air Ministry) and were painted grey, whereas civilian marine engines were usually painted red.

Engines fitted to cars were normally left with aluminium castings bare and the block and head enamelled black; the cooling fan was painted red.

COOLING SYSTEM

Not all early radiators were black; some were painted body colour, as on this AB (far left). The ring for the choke protrudes from the shell. Boyce motometer was a popular and sensible addition, especially for an early Seven without a fan to assist cooling. Sports Sevens during the 1930s used this radiator shell (left), with a separate badge attached by two screws, as used on the 1929/30 touring cars and seen here on a 1934 Speedy – the chromium-plated stone guard rendered it more sporty. A chromium-plated cap screws into the starting handle aperture when it is not in use.

The touring car radiator of the 1930s, here on a 1934 RP saloon, was taller and had the company motif pressed out of the shell. The radiator cap is of Mazak and suffers in old age, but replicas are available. The inner periphery of the shell is correctly painted black.

At first the Seven had merely a radiator for cooling the engine with no provision for a water pump or fan, thermosyphon and the passage of air through the matrix being deemed sufficient. The radiator core had a separate, black-enamelled steel shell, whose winged, nickel-plated Austin motif was attached by a pair of screws. The whole was surmounted by a screw-fitting ebonite cap with threaded brass sleeve. Water was passed between the engine and radiator by 1⅛in three-ply rubber pipes and aluminium manifolds.

At first the radiator cores were of true honeycomb construction but for reasons of economy these were soon substituted for hexagonal film block then, by the mid to late 1920s, interleaf diamond film block.

Some of the owners of the first few hundred cars sold experienced overheating during the summer of 1923 and consequently in October, from car number A1-2145, a belt-driven two-blade fan was fitted. This needed a different timing cover end plate so that a pulley could be attached to drive the fan which was fitted on a separate bracket.

When the Brooklands Super Sports models were standardised with varnished aluminium bodies, the radiator shells on these cars were nickel-plated. The radiator shells used on Gordon England cars and certain other coachbuilt models were made of brass in order to facilitate plating and render them more durable.

In December 1927, from chassis number 50456, the factory altered the position of the radiator drain tap to the offside of the bottom pipe rather than underneath it as before. This change was explained in the service journal as having been made to make the drain tap more accessible.

During August 1928, at around chassis number 67024, a reworked radiator shell came into use; this was slightly higher than the previous type and was nickel-plated for the first year, substituted in 1929

73

by chromium. The factory had begun to abandon outside radiator manufacturers and in the early part of 1929, from car number A8-6850, started exclusively to use components made in-house. These radiators are taller than others and have an increased filler orifice of 1½in diameter rather than the earlier types' 1¼in. The radiator drain tap now protruded through the base of the cowl and the factory fitted a modified front number plate and brackets to allow for this.

In the summer of 1930 the radiator and cowl were altered again to suit the short-scuttle cars then appearing, but the previous design was still used on other models. In July all export cars had a larger cooling fan along with a wider belt and pulleys, and then in October 1931, from chassis number 143000, some export cars began to be supplied with a four-blade fan. A little earlier than this, towards the end of the summer of 1931, the screw-fitting badge was abandoned when the radiator cowl began to be manufactured with the badge incorporated in the top of the pressing.

Ulster, 65, Nippy and Speedy sports cars used the same type of radiator shell fitted to other cars during 1928-29, except that it was chromium-plated. Ulster excepted, these models were also fitted with the bayonet radiator cap that had become standard wear for saloons and tourers from September 1932. This remained in use on the Nippy and Speedy even when the rest of the range had gone over to under-bonnet radiator caps. Ulsters did not have stone guards, except actual racing cars, whereas the 65, Nippy and Speedy models were fitted with a chromium mesh stone guard, so beloved by the 'Promenade Percy' fraternity of the day.

Radiators used on the various military and colonial models are almost impossible to categorise, as they altered from contract to contract. It is safe to say, however, that these models were almost invariably fitted with a supplementary pressed steel internal radiator cowl around the fan, a removable brass gauze filter in the filler neck, and a radiator matrix that altered between contracts. The cowls were either painted or plated according to requirements.

One of the most radical alterations to the visual aspect of the Seven was made with the introduction of the cowled-radiator Ruby model. No longer was the core fitted within the radiator shell; the latter was mounted well in front of the new core, which had an under-bonnet filler in the header tank on the nearside. The shell was of pressed steel with a pot metal badge and a grille with vertical painted pressed steel slats surrounded by a chromium-plated brass surround. The shell was painted body colour and the slats silver. Open two-seaters and vans continued to be manufactured for the rest of the year with the old-style radiators, probably to use up stocks of these and associated body parts.

The radiator cap was moved to the offside in March 1935, from chassis number 219428, in order that the oil and water levels could be attended to at the same time. From chassis number 247309, in 1936, fan pulley bushes were upgraded to 'Oilite' type; export cars have different diameter bushes due to the larger fan spindle fitted to these cars.

The last significant alterations were made in June 1938, at chassis number 286462, when the radiator cowl was replaced by a slightly wider version which had different hinge mountings; the two are not interchangeable. At the same time a relief valve was incorporated into the header tank and this was made available for dealers to fit retrospectively to other cowled radiator cars. Finally, early in 1939 the brass radiator cap was superseded by one of Mazak.

Vintage Sevens were fitted with this ebonite radiator cap, seen on a 1930 AE tourer. The badge is again a separate casting secured by two screws.

EXHAUST SYSTEM

Apart from some of the sports models, the exhaust system of the Austin Seven was always mounted on a cast iron manifold. The same manifold, discharging downwards at the front via a square flange and with a pair of integral inlet ports, was used from the start of production until chassis number 159534 in September 1932, when a one-piece cast iron inlet and exhaust manifold was introduced.

At first, in common with many other cars at that time, the silencer was made up with cast iron ends which fitted into the centre section. The silencer was bolted direct to the chassis by brackets at each end; fresh mounting brackets were available when the centre section became unserviceable and a new cylinder had to be made up. The downpipe and short angled tailpipe were separate from this and a push fit into the ends, the former secured by a long clip pinched together by a pair of nuts and bolts.

Very soon this system of repairable silencer, more suited to expensive cars with large and complicated exhausts, was replaced by a unit construction silencer and tailpipe. This was used until 1929 when, from chassis number 84691, it was decided to extend the tailpipe to the rear. Works coupés had been fitted with this system since their inception, most likely so that their occupants could avoid the disastrous effects of carbon monoxide inhalation, and as saloon bodies were gaining in popularity it was a sensible change. The silencer was moved to underneath the nearside of the rear crossmember, so a fresh downpipe was required in addition to the extended tailpipe.

In September 1932, from chassis number 159534, the downpipe, silencer and tailpipe were all redesigned, although the tailpipe continued to loop over the rear axle. From chassis number 195100 the joint between the silencer and pipe was improved as splits had occurred in extended use.

The next alterations were enforced by the divergence of chassis design during the summer of 1934 into high-frame and low-frame variations. The high-frame cars had a downpipe with combined silencer and tailpipe, while the low-frame cars retained a three-piece exhaust; the tailpipe loop to clear the rear axle was more pronounced on the low-frame cars. Trouble was experienced with this exhaust system drumming against the chassis so, from chassis number 207167, the factory introduced a slightly revised system to eliminate this.

Van exhaust systems began to differ from others during the 1930s, the main alterations taking place at chassis number 166315, during 1935. At the beginning of 1936, from chassis number 240397, the silencer was changed and a longer tailpipe employed.

Factory sports Sevens always had different systems apart from the rather pedestrian but pretty Sports model that was produced at the same time as the Brooklands Super Sports. The latter was endowed with a very grand, fabricated, three-branch manifold that exited through the side of the bonnet, collected into an outside 'lozenge' with screw connection at its rear and from there a single pipe ran back to the tail, normally culminating in a fishtail. Later cars did without the 'lozenge'; instead the manifold unobtrusively merged into a single pipe which ran to the rear via screw connection.

The Ulster Sports had a fabricated three-branch manifold which merged into one pipe as it emerged from the bonnet. A vestigial silencer dropped at the same angle as the front wing, then the pipe ran horizontally above the running board before sweeping up over the rear wing to end in a fishtail. Both silencer and fishtail were manufactured to the specifications required by the Brooklands authorities.

The 65, Nippy and Speedy were equipped with a conventional three-piece system, but this and even the various clips that attached the silencer and tailpipe to the chassis were peculiar to the series.

FUEL SYSTEM

In common with several other cars of the 1920s, the Austin Seven was fitted with a scuttle-mounted petrol tank feeding fuel by gravity to an updraught carburettor.

The fabricated steel tank was of four gallons capacity. Fitted through the bulkhead from the engine bay side, it incorporated a register flange that butted up against this bulkhead and the fixing brackets. The rear location was by a strap passing over the tank and screw ends to brackets on early tanks. Access to the filler was gained by opening the bonnet on the offside. The finish of tanks should be black.

The combined inlet and exhaust manifold was of cast iron. This incorporated a pair of horizontal inlet ports onto which was bolted a cast aluminium manifold, which swept upwards before converging into a single pipe which dropped vertically to a flange. The very first parts list, published during 1922, shows a 22mm F-type Zenith carburettor, devoid of any enriching device for cold starting; this was soon replaced by a Zenith 22 FZ which had a simple hinged air flap to richen the mixture if required. Initially this was operated by a control rod and pull ring which protruded through the radiator cowl, then, by 1924, by a rod and ring which emerged through a bracket screwed to the bottom flange of the dashboard on the passenger side. During the summer of the same year, from chassis number 12025, the petrol tank was modified to incorporate mounting brackets that attached to the bottom flange of the dashboard; at the same time the convolutions of the petrol pipe were altered as there had been some trouble with fracturing. The tap feeding this pipe from the petrol tank was manufactured by Rotherhams.

A Zenith 22 FZ carburettor in bronze was fitted to touring cars. The choke is a simple brass flap with a tiny hole in the centre to allow a sniff of air when closed. Float chamber top, with 'tickler', is retained by a swivelling clip.

The only production model to depart from this specification during this period was the Brooklands Super Sports, which had fabricated manifolding and twin Zenith 30 HK carburettors. These must have been a little over-generous on the carburation, so it is hardly surprising that they were almost immediately replaced by a single 30 MV updraught Solex.

In February 1927, from chassis number 31511, the petrol tank cap was altered from screw to bayonet fitting; with all these scuttle tanks there should be a drop-in brass gauze filter in the neck. During the spring of 1929, in a rather haphazard fashion around chassis number 80000, the bronze Zenith carburettor was superseded by a die-cast one of the same make, type FZB. Different petrol pipe fittings are necessary with this carburettor but it uses the same manifolding. In the autumn the petrol filler was moved to the nearside for convenience when refuelling.

For a little while a fresh series of cars had been taking shape and finally in June 1930 the first of the abbreviated-scuttle saloons appeared – car number B1-301. This saw the end of the old four-gallon tank, to be replaced by a rectangular five-gallon one bolted to the pressed steel bulkhead. This tank has a bayonet-fitting filler cap on the nearside and includes a one-gallon reserve controlled by a two-way tap, which nearly always leaks.

Export cars began to leave the factory with updraught Zenith FZA carburettors fitted with Benjamin-manufactured air cleaners going under the evocative name of 'Air Maze'. At this time on home market cars in the B2 and B3 series the factory briefly tried fitting 22mm Amal updraught carburettors; it is most unusual to come across a car fitted with one of these, and in any case the factory soon reverted to the Zenith FZB. Just prior to this carburation experiment, from car number B2-2500 in September, the petrol tank filler cap was changed back to a screw fitting.

During February 1931 cars of the B3 series underwent a minor redesign of the bodywork which reduced the space available for the petrol tank, so this was changed to a shorter, deeper version of the same capacity. This necessitated a new type of reserve tap and alterations to the petrol pipes for both Zenith or Amal carburettors.

Major redesign work on the Seven during September 1932, from chassis number 159534, resulted in the petrol tank being repositioned at the rear of the car; this tank had an angled filler neck with a chromium-plated bayonet cap emerging at the offside rear. A Smiths electric petrol gauge was at last fitted, with a float-operated electric sender unit set into the top of the tank. No longer would the owner have to carry a spare can or keep opening the bonnet to check the contents of the tank with a dipstick.

At the same time the exhaust and complete inlet manifolding came to be cast as one, the inlet por-

tions now sweeping up above the exhaust into a central inlet flange to which was bolted a sidedraught Zenith 26 VA carburettor. Fuel was supplied by an AC model M mechanical petrol pump bolted to the crankcase below the manifolding, which necessitated a new camshaft with an extra lobe to drive this ancillary. This system suffered some teething troubles with air leaks and by January 1933, from chassis number 167588, the unions on the petrol pump and supply pipes were modified to effect a cure. In June 1934, from chassis number 198747, the AC fuel pump was superseded by the improved type T, and the petrol pipes were modified accordingly.

Export cars were periodically subjected to changes of carburettor: the first sidedraught instrument, in 1932, had Austin parts reference B129 and then in April 1934, from chassis number 193882, this was altered to one with a further improved air filter. Since the early days of motoring English manufacturers had always imagined, rightly or wrongly, that all other countries had extremely dusty, gritty roads and were fond of specifying heavy-duty air cleaners to fend off premature bore and engine wear – something that would lead to a bad reputation and

Zenith 22 FZB carburettor in Mazak fitted to later cars has small differences from the earlier bronze version, such as the clip to retain the float chamber top. Inlet manifold now has a small shoulder around the vacuum pipe to strengthen this area. From the evidence of this totally original 1930 AE tourer, it would appear that manifold and carburettor were black.

The sidedraught Zenith 26 VA carburettor was fitted to a one-piece inlet and exhaust manifold. This is the later, improved air filter used after the spring of 1934, seen on a 1935 AAL Open Road Tourer.

Fuel System

possible increase in guarantee work. Yet they were often oblivious to the fact that similar damage was likely to occur in their own country.

Since the introduction of the Ulster, sports Austin Sevens had been fitted with different carburettors and manifolding from standard road cars. The Sports, or Ulster, in unsupercharged form used a Solex 30 MOV carburettor on a fabricated steel manifold. The blown cars had a Cozette number 4 supercharger with a carburettor of the same make; the updraught inlet manifold from the supercharger was of cast aluminium and incorporated a blow-off valve. The petrol tank was of 5¾ gallons capacity on both models with a 1 gallon reserve controlled by a two-way tap. The fuel filler with bayonet cap was on the nearside.

Supercharged cars were fitted with a rectangular oil tank of about 1 gallon capacity soldered by a peripheral flange to the nearside front face of the petrol tank. On top of the petrol tank was a union to receive the pipe from the pressure pump in the cockpit. Goodness knows why, but the factory considered that it was necessary for these cars to have a degree of pressure in the fuel system, whereas normally-aspirated Sports models relied on gravity feed.

The 65, Nippy and Speedy models used a Zenith 30 VEI downdraught carburettor mounted on a cast aluminium manifold that was in turn bolted to both block and exhaust manifold.

Various changes were made to the general design when the Ruby appeared in July 1934, including slightly different manifolding, a revised accelerator pedal assembly, and a Zenith 26 VA carburettor with interacting choke and throttle. The petrol tank acquired a breather pipe which emerged from the top and a filler neck with rubber hose connector (necessary because the top portion with filler cap

The 65, 75, Nippy (pictured) and Speedy cars had a separate aluminium inlet manifold, and an exhaust manifold to suit, fitted with a Zenith 30 VEI carburettor. The Speedy had a quick-action filler that could have graced a Grand Prix car!

now emerged through the bodywork). In the early part of 1935 the petrol filler neck was further modified to improve its fit through the body.

Tanks and associated steel fittings are finished in black on all cars. Copper pipes and brass unions should be left clean but dull if the correct factory finish is desired.

TRANSMISSION

The Austin Seven, along with many other makes of car large and small, was endowed with a three-speed gearbox during the first decade of its production life.

A single aluminium casting was employed for both casing and bell housing. End covers cast from the same material were attached at the front by screws and at the rear by studs. These end covers held the ball bearing races for the gear shafts and the 'box was dismantled by removing them. The top cover was also an aluminium casting, secured by four studs, and a separate casting incorporating the gate was positioned on top of this by three screws. Some very early cars had the brass oil filler screw cap at the rear of the top cover but this was soon moved to the cover's nearside front.

All gears were straight-cut and there was no synchromesh. Gear ratios on the very first 696cc cars were 14.5:1, 8.25:1 and 4.5:1 with a reverse of 17.0:1; a back axle ratio of 5.5:1 was also tried on some cars. These were altered to 15.7:1, 8.82:1 and 4.9:1 with a reverse of 21.0:1 once the engine had been enlarged.

The clutch was slightly unusual in that the linings were riveted to the flywheel and pressure plate. When worn, the linings would be renewed, but the drive disc could be changed in the event of distortion or wear. This component is attached to its splined drive shaft by six rivets. Although the gear for either mechanical or electric starter is on the clutch cover, the flywheel fitted to the very early cars with no starter is different.

At first the propeller shaft employed a ball joint at the front and therefore it has a curved claw to fit this, as does the third motion shaft of the gearbox. The rear joint was a block and pin universal joint. At around car number A1-2995 the ball joint was replaced by a fabric coupling, which entailed a different third motion shaft with three drilled ears at its end, as did the propeller shaft, to allow the fabric disc to be bolted between them. Gradually, as Seven production got into full swing, small changes were made, in some cases more to pare costs, one suspects, than for reasons relating to engineering improvement – although the ball joint system requires frequent lubrication and seldom received it. This design of propeller shaft remained in use until the introduction of Hardy Spicer couplings in 1933.

Some owners had expressed dissatisfaction with the reverse catch, operated by spring-loaded rod and trigger, so in March 1926, from chassis number 19000, the design was revised; this was not wholly successful so further improvements were made from chassis number 44000. Towards the end of the year, at car number A3-9000, the machining of the gears was differently set to render them more durable and quieter in operation.

During 1927 two small changes were made to the gear lever; from chassis number 41100 it was cranked and, shortly afterwards, from chassis number 41030, it was lengthened. In the autumn of the same year, from chassis number 45016, it was decided to dispense with the rather antiquated speedometer drive – by spring belt from the propeller shaft – and run it from the gearbox. The third motion shaft was altered to accept a spiral bevel gear and the gearbox back plate had provision for the speedometer cable.

By 1928, some cars that had covered high mileages were showing signs of wear in the clutch and third motion shaft splines, so from chassis number 62324 these were lengthened and the third motion shaft bush shortened.

A ball-change gearbox top was introduced as standard from chassis number 99001 in October 1929, but just to complicate matters some earlier chassis were altered at the factory before despatch: these cars have first and reverse on the opposite side from normal ball changes and a differently marked top. Depending upon the type of body to be fitted to the chassis, alternative gear levers were now specified; one for touring and saloon cars and another for coupés and two-seaters. At the same time the felt oil seals at the front and rear of the gearbox were changed to Archimedes screw seals in line with the same modification to the crankshaft oil seal that had taken place just over 1000 chassis numbers earlier.

GEAR RATIOS

1922-32 1st, 15.7:1; 2nd, 8.82:1; 3rd, 4.9:1; reverse, 21.0:1
1932-37 1st, 22.94:1; 2nd, 13.85:1; 3rd, 8.73:1; 4th, 5.25:1; reverse, 29.49:1
1937-39 1st, 22.4:1; 2nd, 13.53:1; 3rd, 8.51:1; 4th, 5.12:1; reverse, 28.8:1
Brooklands Super Sports 1st, 14.5:1; 2nd, 8.17:1; 3rd, 4.4:1; reverse, 19.1:1
Ulster 1st, 13.5:1; 2nd, 7.5:1; 3rd, 5.2:1; reverse, 17.5:1
Ulster S 1st, 12.4:1; 2nd 6.96:1; 3rd 4.89:1; reverse, 16.5:1
65 1st, 22.9:1; 2nd, 13.28:1; 3rd, 8.38:1; 4th, 5.66:1; reverse, 28.18:1
Speedy (to Oct 1934) 1st, 22.94:1; 2nd, 13.85:1; 3rd, 8.73:1; 4th, 5.25:1; reverse, 28.1:1
Speedy (from Oct 1934) 1st, 20.48:1; 2nd, 12.39:1; 3rd, 7.82:1; 4th, 5.25:1; reverse, 26.9:1
Nippy (to Oct 1935) 1st, 21.91:1; 2nd, 13.28:1; 3rd, 8.38:1; 4th 5.62:1; reverse, 29.49:1
Nippy (from Oct 1935) 1st, 20.48:1; 2nd, 12.39:1; 3rd, 7.82:1; 4th, 5.25:1; reverse, 26.9:1

Note Alternative gear sets, usually with closer ratios, were manufactured by the factory for competition work and these found their way onto privately owned cars either to special order or when competition cars were disposed of by the works.

The Ulster sports models, introduced in 1930, had differing ratios from the standard cars and these were 12.4:1, 6.96:1 and 4.9:1 with a reverse of 16.5:1.

During July 1930, from chassis number 113068, a new gearbox top and selector fork were introduced which reverted the gear positions to the original gate pattern, with first gear offside and forward.

A four-speed gearbox, a completely new design having no moving parts interchangeable with the earlier model 'box, was introduced in 1932 at chassis number 159534. Selector rods were now in the top of the 'box and helical gears were used, having ratios of 22.94:1, 13.85:1, 8.73:1 and 5.25:1 with a reverse of 29.49:1. All gear ratios quoted are overall ratios, so these figures are in conjunction with the lower axle ratio of 5.25:1 introduced at the same time.

These early four-speed 'boxes suffered from rather flimsy gear levers at the point where they entered the top, so this component was strengthened from chassis number 175450. Shortly after this, during August 1933 from chassis number 179368, synchromesh was introduced on third and top gears; henceforth some of the internals used were the same as the Seven's larger relation, the Austin Ten. Synchromesh for second gear did not appear until a year later, from chassis number 198678, and was used on all models except vans, which only had this refinement on the top two gears until they too acquired it on second gear towards the end of 1935, from chassis number 235580. Contrary to what one might assume, the earlier crash 'box is considerably more robust.

The next significant alteration to the transmission was made during August 1936, from chassis number 249701, when the clutch was brought up to date by abandoning the lined flywheel and pressure plate for a lined friction plate supplied by Borg and Beck. A Newton Bennet clutch became an alternative to the Borg and Beck in early 1938 from chassis number

TRANSMISSION

282122 and then completely replaced it during July from chassis number 289251.

Fresh gearbox front and rear covers were necessary when the oil seals were altered to leather around engine number 279446 in November 1937. Second gear on the four-speed 'box was prone to become noisy with age so at the beginning of 1938, from chassis number 281995, the third gear shaft and sleeve were modified to prevent wear occurring on the overrun.

In November 1937 the rear axle was changed to a higher ratio of 5.125:1 and so, although the gears remained the same, overall ratios became 22.4:1, 13.53:1, 8.51:1 and 5.125:1 with a reverse of 28.8:1.

The 65 and Nippy, fitted with a rear axle ratio of 5.625:1, had overall gear ratios of 21.9:1, 13.28:1, 8.38:1 and 5.625:1 with a reverse of 29.49:1. Later cars, fitted with a 5.25:1 rear axle, had ratios of 20.48:1, 12.39:1, 7.82:1 and 5.25:1 with a reverse of 28.1:1. The Speedy used a 5.25:1 rear axle and, at first, a gearbox with ratios of 22.94:1, 13.85:1, 8.73:1 and 5.25:1 with a reverse of 28.9:1; the last cars of this small series had the same ratios as the later Nippy models.

Aluminium castings should be left bare and steel fittings, such as the gear lever, should be black.

WHEELS & TYRES

From the very beginning the Seven was fitted with a novel design of wheel that allows its removal without fully undoing the wheel nuts. The wheel is located by three dowels standing proud of the brake drum; the three wheel studs and their nuts serve only to clamp the wheel securely to the drum. In between each locating hole and that used for the stud is a radiused slot which enables the wheel, once the nuts have been slackened, to be withdrawn by rotating it until the larger dowel holes line up with the wheel nuts. This system was used until August 1938 when, from chassis number 287828, the three wheel studs alone were used to attach the wheels.

Quite apart from the foregoing, there were many variations of both hub and rim. Open-centre hubs were employed with the 6in brakes and until 1925 a beaded-edge 700-80 rim carrying a Dunlop 26×3in tyre was used, except for vans which had 26×3½in.

In search of greater comfort, well-based rims were introduced in February 1925, at car number A1-8250, and fitted with lower pressure 3.50×19in Dunlop balloon tyres, which had been available upon special request since early 1924. There were two types of rim: a normal rounded-profile, rolled edge, and for a short time another with a flatter, slightly concave edge.

Brooklands wheels with alternative spoking arrangements were catalogued, normally to be fitted to the Brooklands Super Sports model. These had 38 spokes instead of the 36 found on all other models.

In September 1926, at car number A3-3727, an increase in wheel centre size was necessary owing to the enlargement of the brakes, and at the same time the open centres were abandoned in favour of raised, flat-ended ones stamped with the Austin motif. An increase in spoke diameter to 8 gauge was considered prudent owing to the steadily increasing weight of

To begin with the Seven was fitted with 26×3 beaded-edge tyres on wire wheels (above), the open centres of which exposed the screw-on aluminium cover for the wheel bearings. The rarer variation of the well-base wheel rim (kft) used in 1926/27; the more commonly found wheel has a conventional rolled edge. The valve cap is typical of the period.

From November 1931 19in wheels with 'Staybrite' centres were fitted (left), as shown on a 1934 RP saloon. Home-market cars in the 'jewel' series were fitted with these 17in diameter wheels (centre), here on a 1935 AAL Open Road Tourer. From July 1936 the wheel centres were enlarged (right), as seen on an AC Pearl.

bodies, and this was effected in June 1928 from chassis number 63806. This style of wheel, with 18 outer and 18 inner spokes, was used until June 1931, when a further increase in spoke diameter and wheel centres of a thicker gauge steel were specified from chassis number 138350.

During November 1931, at around chassis number 144380, a new style of wheel appeared. The size of rim and tyre remained the same, but now the wheel was laced with 12 outer and 24 inner spokes and the hub had a pressed-in stainless steel wheel centre. It has been said that standard, as opposed to de luxe, models left the factory with this centre painted black, but examination of original factory literature shows that this is not strictly true, although black centres were available.

When the Ruby appeared both it and the Pearl were fitted with 17in wheels wearing 4.00-section tyres, while the open and sports models continued to use 19in wheels with 3.50-section Dunlops. Export Ruby and Pearl models had 16in wheels with 4.75-section tyres, the other models having 18in wheels with 4.00-section tyres. These 16in wheels could also be obtained by special request on home market sports models.

In July 1936, at chassis number 249701, the wheel centres were enlarged and the 'Staybrite' stainless steel wheel centres became a larger diameter, enveloping pressing rather than just an insert.

In 1938, as previously mentioned, the wheels were redesigned and at the same time reduced in diameter to 17in with 4.00-section tyres, except for export cars which had 16in rims with 4.75-section tyres. Dunlop ELP (Extra Low Pressure) tyres continued to be fitted as standard as they had been for the previous three years or so.

If you come across any 19in or 17in Austin Seven wheels with little projecting 'mushrooms' secured to their centres by six rivets, these were fitted to cars supplied to the military and some chassis destined for countries with difficult conditions, such as India. Handles were supplied for fitting over these projections in order that the vehicle could be lifted over impassable terrain. They were fitted with larger section tyres – 4.00×19in or 4.75×17in.

Vans on the low chassis were as a rule supplied with 18in wheels and 4.00-section Dunlop tyres.

The standard finish for wheels was black stove enamel, with the exception of the aforementioned wheel centres.

REAR AXLE

Throughout its life the Austin Seven was fitted with a rear axle which, although the basic concept remained the same, underwent a series of changes and improvements.

It is of the torque tube type, with spur gear differential and halfshafts having integral pinions and a taper with Woodruff key location for the hubs. The axle casing is made of three bolted-up sections – the differential carrier and a pair of axle tubes. Both differential and pinion shaft run on ball bearings with the exception of the rear pinion bearing, which has a roller race. Meshing of the pinion with the crown wheel is adjusted by shim packing of the torque tube flange and the requisite end-float obtained by a screw cup and lock nut on the pinion shaft flange. In July 1924, from car number A1-5047, the size of the threaded portion of these components was reduced from 1¾in to 1½in. From the beginning of production until 1929, at chassis number 86766, the torque tube was a flange fitting, secured by six studs to the differential casing.

Tapered-bore hub bearings were employed until

As well as oiling charts given out in an envelope to owners, the factory produced large garage charts; this one, factory publication number 1097G, dates from 1934.

1926 when, from around car number A3-6000, they were superseded by parallel-bore components. This was mainly for reasons of economy of manufacture because, in spite of the taper races having some advantages, they were a non-standard part.

In June 1928, from chassis number 64801, some redesign of the differential occurred with the result that the differential carrier (or differential case as the factory confusingly called it) came to be manufactured in equal rather than asymmetrical halves. The crown wheel to fit this carrier is machined for different mounting bolts although the ratio remained unaltered at 9/44; a replacement crown wheel must therefore be of the correct type.

Between chassis numbers 84000 and 86750, in May 1929, a new design of screw-in torque tube began to be fitted. This allowed the mesh of the crown wheel and pinion to be adjusted without parting the torque tube from the remainder of the axle assembly, as had been the case with the previous shimmed pattern. At the same time the pinion shaft was enlarged from ⅞in to 1in diameter and the roller bearing at the front of this shaft was changed to a ball bearing.

AXLE RATIOS

Early tourers (696cc engine): 10/55, final drive ratio 5.5:1
Tourers, saloons and vans, 1923-32: 9/44, final drive ratio 4.9:1
Brooklands Super Sports, 1923-26: 10/44 or 9/44, final drive ratio 4.4:1 or 4.9:1
Sports (Ulster), unsupercharged: 9/44 or 9/47, final drive ratio 4.9:1 or 5.2:1
   Supercharged: 9/44 or 9/51, final drive ratio 4.9:1 or 5.6:1
Milk delivery and military: 9/51, final drive ratio 5.6:1
Cars and vans, 1932-37: 8/42, final drive ratio 5.2:1
65, Nippy and Speedy: 8/45, final drive ratio 5.6:1
Cars and vans, 1937-39: 8/41, final drive ratio 5.1:1

Note The pre-production sales leaflet stated that the Austin Seven tourer had a final drive ratio of 4.5:1. Whether or not a few prototypes were made with this ratio is open to conjecture, but the surviving car in the Science Museum in London might provide the answer.

This axle is the type fitted to almost all Ulsters – certainly all cars having the lower axle ratios of 5.6:1 or 5.22:1 – and is immediately recognisable by the enlarged differential casing to accommodate the oversize crown wheel, and by the webbed axle tubes. The 5.6:1 ratio axle was not peculiar to the Ulster: it was used on Milk Delivery vehicles and most military models too.

Crown wheel and pinion gear angles were altered in the summer of 1930 from chassis number 115234 as some concern was caused by an increasing occurrence of gear wheel troubles since the introduction of the new axle just over a year before. Shortly afterwards, in September, between chassis numbers 116799 and 116881, the factory began to experiment with another design of torque tube – the screw-in tube had insufficient pinion bearing location which could lead to excessive end float developing in the pinion shaft and resultant axle failure.

Once more a flange fitting was employed but the rear pinion bearing was mounted in the axle case and the shims for adjusting the mesh of the crown wheel and pinion were made so that it was unnecessary to remove the torque tube when carrying out this procedure. The factory clearly was not happy with this design as it reverted to supplying cars with the previous screw-in torque tube rear axle between chassis numbers 116882 and 118500.

Finally it was decided, in September, to put yet another improved axle into production and this followed after chassis number 118500. Once again it had a flange-fitting torque tube retained by six set screws. The pinion shaft had a pair of rear ball bearings positioned by distance pieces and retained by a nut; end float was not adjustable. Correct pinion mesh was obtained by shimming the flanges, the feature of easily changeable shims being retained from the previous short-lived design.

Various minor improvements were made to this axle during the first few months of its manufacture, including from November 1930 an increase in the size of the oil filler aperture, and Archimedes principle oil seals machined into the crown wheel adjustment collars from chassis number 122030 in the same month. In January 1931 the pinion bearing retaining nut became secured by a tab washer rather than a grub screw, and the differential carrier shafts were increased in diameter. Teething troubles were gradually eradicated; for example, to avoid loosening pinion shaft bearings the factory first changed the bearing retaining nut to a left-hand thread, from chassis number 133178, then the outer distance piece received the same modification, from chassis number 133313.

The three-piece axle casing that had survived from the outset was finally replaced by a two-piece unit from chassis number 159534, in September 1932. This is sometimes referred to as the 'D' axle owing to its sectional shape; the offside tube was a stamping while the nearside tube was unaltered from the previous axle. Half shafts from then on had a shoulder of marginally increased diameter for the last few inches of their length nearest to the pinion.

Although not strictly part of the rear axle, mention should be made here of the torque tube socket fitted to all cars. The front of the torque tube is an open sphere from which the flange for the propeller shaft protrudes; this sphere articulates within a socket, at the base of which is a tube projecting downwards. This is slotted and accepts a ball flange which itself is riveted (apart from some later cars which had bolted flanges) to the chassis rear crossmember; the ball is held within the tube by a pair of cups secured with a nut and locknut.

Visually there are four types of torque tube. The first, used until 1930, has a single long locking bolt passing through the top of the tube to secure the torque tube retaining ring. The second has a pair of locking bolts at three and six o'clock through a small flange on the socket. The third has a longer, stepped tube and is used on the low-chassis cars. The fourth is bolted to the chassis with a Silentbloc bush. All four types should be kept well greased.

After five years using ball bearings for the pinion the factory decided that the previous design gave better support and so from chassis number 205849, in November 1934, the rear of the pinion shaft was increased in diameter and a roller bearing was once again fitted.

Between chassis numbers 239342 and 248457, in the spring of 1936, just over 500 cars were fitted with the new Silentbloc bush mounted torque tube socket; this became standardised from chassis number 254451. At the same time, during April, the configuration of the pinion bearings was slightly altered to give some pre-load.

In 1937, from chassis number 276596, the factory fitted a pair of felt oil seals on the pinion shaft at either end of the torque tube; shortly afterwards, from chassis number 276998, the oil filler was moved to the offside and increased in size. At the close of the year, from chassis number 281472, the crown wheel acquired 41 teeth, necessitating a new pinion.

The Seven underwent several changes in axle ratio and crown wheel and pinion sets over the years and obviously it is imperative to fit a correctly matching pair. There is no use delving around in a box of spares and coming up with a component with just the correct number of teeth. The crown wheel and pinion must match.

FRONT SUSPENSION

The front suspension of the Seven is by means of a single transverse leaf spring supporting a forged axle beam which is controlled by radius arms running back to a ball joint in the centre of the front crossmember and a single transverse shock absorber. At first the shock absorber was not fitted, so early models, until car number A1-3145, therefore had a front axle with a plain beam devoid of any provision to attach a shock absorber. The changeover period was a little haphazard but by car number A1-4052 all vehicles were fitted with the front axle that was to remain standard wear, except for sports models, until 1936. At first Austin fitted proprietary shock absorbers by Hartford but by the summer of 1925, from around chassis number 10653, it had tooled up to make its own.

At the car's inception the front spring had five leaves of 1½in width and 3½in camber, but the camber was altered to 2⅞in in 1926 and then during 1929, from chassis number 90030, the number of leaves was increased to eight while the camber became 3in. Suspension fitted to Military and Colonial models had springs with greater camber in order to increase ground clearance. From June 1935 a more robust spring was fitted to export chassis. By

Due to the fact that this 1930 AE tourer has never been dismantled or restored, it appears that the front axle, suspension and steering were dipped in paint after assembly.

Front Suspension

Beginning with the Ulster, sports Sevens were fitted with this bowed front axle, seen on a 1936 AEB Nippy.

These views show the front axle, brakes and some of the steering assembly on a 1934 RP saloon (right) and a 1934 ARQ Ruby (below right). The brakes on the RP saloon are semi-Girlings, which are incorrect for the period – but the owner has fitted them to improve braking.

1936 seven leaves were employed and during 1937, from chassis number 278708, vulcanised fibre packing was used rather than the hardwood that had been employed before.

Towards the end of 1926 both front and rear axles, which up until that time had been fitted with tapered-bore wheel bearings, began to be equipped with parallel-bore bearings, at around car number A6-6000. This entailed new front swivel axles and rear axle tubes but, as it is possible to fit these to earlier cars, it is not unusual to find them so modified if the original parts have become worn out.

Shock absorbers underwent a gradual series of minor improvements. Bushing of both front and rear arms was strengthened in March 1928, from chassis number 53000. In September 1929, from chassis number 95541, the rubber bush walls were thickened. During April 1934, from chassis number 193954, more substantial front shock absorbers and bracketry were fitted to export models. July 1937 saw the aluminium links between axle and shock absorber lengthened from chassis number 276213 then, during March 1938 at around chassis number 284581, the rubber bushes carried within these links became flanged (until chassis number 56275, in 1928, these bushes had been of hardwood).

Sports Austin Sevens manufactured between 1929–36 were fitted with a variation of the normal front axle that was forged into a downward curve between the shock absorber mounting holes. This was to allow the reverse camber spring adequate movement; springs fitted to these cars had seven leaves, 1½in width and reverse camber of 2⅛in, and required slightly different spring packing and clips.

Very passable replicas of the above axles have been made from standard ones, but without very careful forging the curve is not true and the axle tends to end up a fraction shorter than the original; the front track should measure 3ft 4in. Standard axles of this period have a part reference number BL 1 while the sports component has the number BL84 (part number 9C 135).

In August 1936, from chassis 249701, the front axle was provided with holes to attach the stronger radius arms required by the introduction of Girling brakes. A dropped version of this axle was fitted to works 'Grasshopper' cars prior to the adoption of Girling brakes, due to the heavy radius rods used on these cars.

Axles, radius arms and shock absorbers should be finished in black and the springs left bare; sometimes the latter had the part number stencilled on them in white (eg, 1A-5437 on a Ruby). Springs on the sports models were normally bound with cord or wrapped in tape; both methods were employed by the factory.

The front axle of the Seven is rather too light in construction for the weight that the cars had attained by the early 1930s and as a result the kingpin eyes can become badly worn. It is possible to have this rectified by a specialist so do not discard an axle that has distorted eyes or is bent.

There are various U bolts – the factory referred to them as spring clips – to match the different springs fitted to the various models and it is essential to use the correct ones. Friction discs within the shock absorbers must be in good condition; before the early 1930s the centre disc was thicker than the outer two but later cars have uniform discs.

83

REAR SUSPENSION

Suspension at the rear is by a pair of quarter-elliptic springs and, apart from early cars up until car number A1-3145, the ride is controlled by friction shock absorbers. Hartfords were used until 1925, when Austin began to manufacture and fit its own components from around chassis number 10653. Rear spring pins, incidentally, are dissimilar with the two varieties of shock absorber.

Until 1926 the rear springs had six leaves with a camber of 8in, but from the late summer, at car number A3-3361, the number of leaves was increased to seven with an 8⅛in camber; the width of the spring leaves was 1½in. The large bolts that locate the spring were supplied in slightly greater length when the second type of spring was introduced. Springs destined for use on vehicles supplied to the War Office and for colonial service could have special camber or strength depending upon the individual requirements of the contract in hand.

Sports variants had six-leaf springs with no appreciable camber and these could be bound either with cord or tape. The aluminium connecting links for the rear shock absorbers were longer on these cars; at chassis number 96867 during September 1929 the bushes changed from wood to rubber.

During August 1930, from chassis number 114871, it was deemed necessary to strengthen the rear springs on saloons due to increasing weight. In 1934, at chassis number 190001, the factory began to specify a stiffer spring on the offside to negate the effect of the weight bias if the driver was the only occupant. In the spring of 1938, at around chassis number 284581, the rubber inserts in the shock absorber links became flanged, at front and rear.

Shock absorbers should be finished in black and the springs left in bare metal. Regular attention with the grease gun will help to delay the rather tricky job of replacing rear springs and bushes.

Rear shock absorbers were similar throughout the life of the Austin Seven. This is a 1934 ARQ Ruby, with the low chassis frame extending upwards over the rear axle.

STEERING

The conventional worm and wheel steering box has a cast aluminium casing with steel tube column. The box is bolted to the chassis and the brake fulcrum pin passes through them both. A one-piece gear wheel and drop arm connects, by spring-loaded ball joint, to the drag link which in turn articulates with the steering arm via another ball joint. The track rod is only adjustable at one end on early cars; the track rod ends (or jaws, as the factory termed them) connect to the steering arms by pins which swivel.

The three-spoke steering wheel is secured to the steering column by a keyway and clamp bolt; four-spoke wheels were fitted to early cars although strangely they were still listed as current by the factory as late as 1927. The four-spoke wheel has very little dish, but the three-spoke variety is very deeply dished; both are coated with black celluloid. The three-spoke wheel was supplied with a rim incorporating either inner corrugations for the fingers or spiral undulations; the latter type is far more rare and was used mainly on cars from 1927, until dished wheels were discontinued in July 1928. Before chassis number 12328, in September 1925, the horn button was on the dashboard, although the pre-production sales leaflet stated that an electric horn operated from the steering wheel was fitted.

A pair of nickel-plated control levers marked GAS and IGNITION swing across a circular plate on the centre of the wheel and control the throttle and ignition timing. The GAS lever has a domed centre on cars with a dashboard-mounted horn button. Small spring-loaded friction blocks let into the reverse of these levers bear on the V rim of the plate; a pair of tubes rotate, one inside the other, through the steering column and connect via tiny levers at the base of the steering box to throttle and magneto.

The new design of steering wheel with much less pronounced dish was introduced at chassis number 65442. The method of locating it onto the column was altered to splines and at the same time the splines at the opposite end that located the worm were deepened. A felt bush was now used in the top of the column to block the path of any oil seeping up.

When the short-scuttle saloon bodies appeared in June 1930 a different rake was required for the steering column, so the box casting was re-patterned to allow for this. The next improvements were made in the spring of 1931 when the bushing in the box was altered from chassis number 130860. During October of the same year, from chassis number 143183, the steering box became cast in steel and was attached to the chassis by set screws; then from September 1933, at around chassis number 180000, the steering box was once again cast in aluminium and bolted to the chassis frame.

Towards the end of 1932 the quadrant and levers in the centre of the steering wheel began to be fitted with a Bakelite cover and an enlarged horn button, until the appearance of the Ruby in 1934. A new steering wheel with bulbous boss was then adopted; the hollow within accepted the switchgear for the trafficators that were now fitted. A Bakelite cover with chromium horn push covered the centre and a chromium lever controlled the trafficators.

At the same time as the standard chassis were being made, the special sports variants of the Seven were also in production; these often have a greater rake to the steering column to allow a lower body line. During the 1920s three choices of steering rake were available – 49, 41 or 37½ degrees from chassis frame to column.

The Ulster had a regular production steering wheel while the 65, Nippy and Speedy were normally fitted with a four-spoke Bluemels wheel, coated in black celluloid and with fine ribbing on the rim between each spoke. Later Speedy and Nippy models often had an Ashby three-spoke sports wheel to allow the horn and trafficator assembly from the Ruby to be used.

In October 1936, during the chassis number sequence 256000 to 257000, a completely new steering box was introduced, known at the factory as the 'hourglass' type. It was of worm and sector type and the drop arm was no longer integral with its shaft, allowing removal for adjustment without dismantling the box. This box had a greater range of adjustment than the previous model and the worm was supported by ball races. While this changeover was taking place, it was found necessary to strengthen the mountings along with the use of a fresh fulcrum pin which still passed through the mounting of the box – this was carried out from chassis number 256806. Just before this, from chassis number 256335, the drag link began to be fitted with additional grease nipples and then at the beginning of 1937, from chassis number 265295, the steering box cover was modified to accept self-lubricating bushes.

The finish of components is black, as with all other painted parts on the Austin Seven chassis, except that the steering box casting, where aluminium is employed, should be left unpainted.

BRAKES

When the Austin Seven was conceived, four-wheel braking, even if uncoupled with the hand brake operating the front and the foot brake the rear, was still something of a novelty, especially for such a tiny car.

Drums of 6in diameter with a shoe width of 1in were initially thought sufficient. Those at the rear are operated by a pair of cables attached to levers on a cross-shaft, which is actuated by the brake pedal linked to it by an adjustable rod. The front brakes are operated by a single cable, looped around an adjustable pulley and bracket assembly that is bolted to the hand brake lever, the ends then running forward to the cam levers on the back plates. These cam levers, front and rear, have ball ends onto which attach sockets at the extremity of each cable. The sockets are constructed of two pieces, male and female, and are secured by a through bolt which also keeps in place a small spike, the purpose of which is to splay the cable and prevent it from pulling out of the socket assembly.

After the first few cars had been made, the back plates were increased in diameter slightly to reduce the ingress of water and small stones. During September 1926, at car number A3-3727, the diameter of the brakes was enlarged to 7in but the lining width remained 1in. At the same time the brake cam levers were lengthened and the shoe pivot pins were riveted rather than having a threaded shaft and nut. In the summer of 1928, at around chassis number 64643, the brake pedal was slightly modified to give the car a better turning circle. A further small improvement was made towards the end of June 1929, at chassis number 89501, when the hand brake trigger and spring assembly were altered to prevent unintentional disengagement.

By 1930 the uncoupled brakes were looking distinctly old-fashioned and in July, from around chassis number 113000, the system was redesigned so that both hand and foot brake operated on all four wheels. The front brake cable pulley was now mounted to a lever clamped onto the cross-shaft, and the hand brake lever bore, via a roller, on an adjacent rocker. Hand brake adjustment is by means of a bolt and a wing nut between rocker and lever. In order to equalise front and rear brakes, the lever is set in the desired position on the cross-shaft.

Pedal adjustment remained as before, with a wing nut on the foot pedal adjusting rod; the rod and catch for the hand brake was altered from Whitworth to AB thread from chassis number 137770 and shortly after, at chassis number 140300 the brake cams received greaseless bushing.

During 1932 the factory experimented with different brake cam levers and then decided to revert to the type introduced in 1926; alternative levers were fitted between chassis numbers 148625 and 159844. Shoe width was increased to 1¼in in the autumn, from chassis number 159534.

The 65, Nippy and Speedy variants used differing hand and foot brake levers, parts reference numbers BK 87 and BK 93, to allow for the different configuration of these models.

Briefly, during the summer of 1934, the factory tried using alternative brake lining material but this was soon abandoned. Braking performance – or rather the lack of it – was obviously a cause for some concern as by the summer of 1935, from chassis number 226847, further modifications were made, this time to the pivots and shoes to try to improve matters by increasing the leverage; another type of lining material was tried at the same time.

May 1936 saw the brake drums changed to cast iron in place of pressed steel, this commencing at chassis number 246176. By the end of the month, from chassis number 246835, the aluminium brake shoes were replaced by fabricated steel components.

Continuing with its attempts to improve the braking and nullify the effects of the Seven's weight increase since the 7in brakes first appeared in 1926, the factory resorted to using a semi-Girling system from the beginning of August 1936. This meant that each brake was separately adjustable at the back plate but cable operation was still retained.

During July 1937, from chassis number 276613, the brake shoes became interchangeable so that they could be swapped over, if desired, to equalise wear. In October, at chassis number 279753, the front brake cable swivel was strengthened.

The final alterations to the braking system were made in July 1938, intermittently from chassis number 286440 and then to all cars from chassis number 286571. They consisted of the adoption of virtually the entire rear brake set-up from the Big Seven so that the rear brakes now had the benefit of full Girling operation.

Much criticism has been levelled at the brakes on Austin Sevens and it is true that they are not the car's best feature. Remember, however, that when the Seven first appeared on the comparatively deserted roads of the 1920s, brakes were not so important as they are today. The earlier, lighter models were far better at stopping, even with uncoupled brakes, but as the bodywork became heavier naturally the cars became less easy to retard, hence the multiplicity of alterations throughout the 1930s which were not assisted by the fact that coupled brakes never achieved satisfactory front and rear compensation.

In order to maximise braking efficiency it is important to check brake shoes periodically for oil as seals are prone to leak, and incorrect grease in the front hubs can find its way into the wrong places. Oil on the shoes can result in juddering, grabbing or in severe cases hardly any retardation at all. Another cause of juddering or uneven braking can be worn linkages, suspension points or cam bushes, while cams also seize up.

If you have just purchased a car and you find it fitted with those ghastly proprietary cable adjusters, I would recommend throwing them away and instead, if the cables are stretched, treating the car to a new set. Regular maintenance, oiling and greasing all necessary points and replacing worn components can go a long way to quashing the Seven's reputation for poor brakes.

At first the brakes were of 6in diameter. Although not visible here, the front axle on this 1923 AB tourer has no provision for a shock absorber.

During 1926 7in brakes were introduced. The tailored leather spring gaiters on this 1926 R saloon were an addition that could be fitted to any car, and are still obtainable.

ELECTRICS & LAMPS

Before the Austin Seven entered production several sources of electrical accessories – such as Smiths headlamps – had been investigated by the factory. But as production got underway in 1922, the first factory parts list shows the following as original equipment.

The magneto was a Scintilla type M4, and everything else was supplied by C A Vandervell apart from the horn and the dashboard-mounted button for it, which were both by Benjamin. The largest sidelamps made by Vandervell, the bell-shaped type FWA, did very nicely as headlamps and a single type CTS gave a red light to the rear and illuminated the number plate. The battery, also supplied by CAV and located under the driver's seat, was kept charged by a type DF dynamo with an output of 5 amps in conjunction with a type E cut-out and type 12 switchboard with black-faced ammeter.

One of the first alterations to the above specification was to fit a BLIC magneto and then, as bell-shaped headlamps were going out of fashion, a change was made to CAV type ACPs – again in fact a large sidelamp. As an alternative a small headlamp made by Howes and Burley was used.

At the end of 1923 electric starting became standard and again CAV components were used for both starter and switch. To cope with the extra demands now put upon it, the battery was uprated to a CAV type 6XW2, and a CAV type DFL dynamo, with a slightly longer body than before and an output of 8 amps, came into use.

In 1924 yet another set of CAV lamps appeared on the Seven – type AP with screw-fitting rim at the front, and a more modern type AT at the rear.

During the first months of 1926 the automatic fitting of BLIC magnetos ceased and it became up to customers, if they so desired, to specify the make they required. It is doubtful if many of them actually took advantage of this option, so the factory had licence to fit whatever was readily available at the right price. Besides BLIC, the magneto could hereafter come from BTH, Lucas, SEV, Watford or ML. If you find a car with the magneto missing but retaining the straps, it may be of interest that those for the BLIC are shorter than for the others.

In the same year the CAV company became part of Lucas and a change was made to Lucas electrical equipment. Up to and during this period the supply of CAV lamps began to dry up, so the factory used on a few cars the similar sized flush-rim lamp by Howes and Burley. These very probably were stock left over at the factory and it was convenient to use them at this time.

Shortly before this the Benjamin horns had been discontinued and a choice of two other makes began to be fitted; the most common is a Rist, the other a Sygnol, and by this time the dashboard horn push had been replaced by one in the centre of the steering wheel. The mounting bracket for these horns differs from the earlier one in that it is T-shaped in section and the mounting holes will not pick up on a Benjamin horn.

The changeover from CAV to Lucas equipment did not happen overnight so it would be misleading to give a particular chassis number. It is known that many cars were produced with all or partial Lucas electrics before the factory noted that from chassis number 25140 the Lucas SM3 dashboard switch panel was standardised. This had a pair of switches below an ammeter, marked off/magneto/dynamo and off/side/head, with a pair of sockets for ancillaries between them. The cut-out fitted in conjunction with this was a CF1, having a separate fuse cover. The dynamo remained a DFL and the battery was a 6TW9. Starting was by Lucas M35 starter motor operated by a Lucas switch; the CAV switch has two mounting holes and the Lucas three, so the starter mounting case should be of the correct type. Headlamps, still for the moment mounted on the scuttle, became Lucas model R 515 and the single rear lamp a Lucas AT 201.

At this time a windscreen wiper became standard except on the Ulster when it went into production in 1930. Although at first it was not electrically powered, I have included all wipers here because

By mid-1926 Lucas electrical equipment was being used on the Seven; covers of the fuse box and cut-out are anodised brass. Also clearly visible on this R saloon are the levers for operating the advance and retard mechanism on the magneto.

87

they were made by Lucas, and obviously their adoption was not wholly unconnected with the change to this make of electrical equipment. The wiper first used was a vacuum model 25, activated via a black rubber pipe from the inlet manifold, which from now on was drilled and fitted with a short length of brass pipe.

It was at this time that the factory experimented with moving the headlamps forward onto the front wings. Around 2000 cars were produced in this form, with R40M headlamps, over the winter of 1927/28 from car numbers A3-4572 to A3-6754 before the headlamps reverted to their former position when their legality was questioned.

In August 1928, from chassis number 67024, the factory permanently altered the headlamp mounting position to the front wings. At first this caused a minor furore as owners were intimidated by some police forces who were unaware that the lighting laws, especially those concerning the positioning of lamps within the outer extremities of the bodywork, had been very recently reviewed. The Austin factory was well aware of this situation and consequently issued a statement for owners to show to the police should the need arise. These headlamps, R47s, differed from the previous ones in that there was a sidelamp bulb incorporated in the reflector.

The following month saw the introduction of coil ignition, which led to a special version of the CAV DFL dynamo being produced by Lucas. It had a casting at one end to house a right-angle drive and the distributor; a new aluminium casting to house the dynamo was also necessary due to the latter having a three-bolt flange fixing rather than the clamp type that had been used until then. Ignition was by CAV distributor and Lucas coil, although any product carrying the CAV name was by then manufactured under the aegis of Lucas.

The switch panel for lights and ignition had been changed to the SM 5 model, this having an ignition warning lamp in place of the pair of plug sockets. The switch panels fitted to Austins almost invariably have a day, month and year scratched onto the plate at the rear; although this is invisible when the instrument is in the dashboard, one having a date immediately preceding that of the registration of the car is a plus point.

From December 1928, at around chassis number 74000, an alternative ML ignition coil and distributor began to be fitted to some cars, and continued in use until 1931. These components are easily recognised: the distributor body is more squat than the Lucas and the coil is very short with an external resistance bobbin on the top and a flanged base with mounting holes.

The old 'bacon slicer' starter was done away with in October 1929 and from chassis number 99001 a two-bolt, flange-fitting starter was introduced which

still mounted rearwards on top of the bellhousing but was considerably less bulky. This was due in part to the switch, a two-bolt flange-fitting version of the ST8, being mounted on the transmission tunnel. Another modification made to the Seven at the time was the repositioning of the petrol filler, which meant moving the horn mounting to clear it.

One of the problems with the earlier type of Lucas vacuum wiper was that the arm gradually dropped into the vertical position when it was not in use, interfering with the driver's vision. To overcome this the improved type 30 was developed, incorporating a sliding on/off switch behind the regulating screw, which on the previous model had to be slackened to turn off the wiper. This was introduced during the autumn of 1930.

In the summer of 1931 dipping headlamps made their appearance on the Seven. These were Lucas R47s, to which the factory fitted and recommended

Even after the change to Lucas, a CAV dynamo continued to be used (top). The magneto fitted to this 1926 R saloon is a BTH. The Rist horn (above) that was fitted to many models, seen on a 1932 Swallow saloon.

ELECTRICS & LAMPS

In real racing fashion the rear of the dashboard on the Speedy is fully accessible and makes an impressive sight.

The Nippy has its cut-out, fuse box and junction box mounted at the rear of the engine bay.

The later version of the Lucas vacuum wiper, type 30G (above, top) looks very similar to the earlier one. Removed from the car (above, centre), a 1934 RP saloon, the 'hidden' side shows fixing screws and spindle, and also the pipe onto which attaches the rubber hose leading from the inlet manifold; the unit is sitting on a running board with the correct rubber moulding. A rather endearing electric wiper motor (above, bottom) was fitted to the Ruby and Pearl as well as the tourer.

89

Original Austin Seven

The three types of scuttle-mounted headlamp (left-hand column). Bell-shaped CAV lamp (top) was fitted only to early cars, seen on a 1923 AB tourer. CAV lamp with screw-fitting rim (centre) used between 1924-26 and shown on a 1924 AC tourer; curved mounting bracket and armoured cable are visible. Lucas R515 lamp (bottom) on a 1926 AD tourer.

Lucas R47 headlamp (right) with sidelight bulb above the main bulb, on a 1930 RK saloon.

Lucas LB 130P headlamp (right) with sidelight bulb below the main bulb, and with correct diffuser glass, on 1934 RP saloon.

A very late box saloon (1934 RP) with Lucas LB 130P headlamps (below); the different appearance of each lamp is due to the one on the right having the dipping solenoid and a more recessed glass.

ELECTRICS & LAMPS

A post-Ulster sports model headlamp with distinctive ribbed glass, seen on a 1934 Speedy.

The headlamp and sidelight of the first series of 'jewel' cars. Normally the headlamp would have a chromium-plated shell, but this 1934 ARQ Ruby is correctly finished like this as it is the cheaper model.

Lucas LBD 131 headlamp and LBD 109 sidelight fitted for the 1938 model year, seen on a 1937 ACA Pearl.

Lucas Graves double-filament bulbs. The dipping switch, a Lucas 9/CS, was mounted on the steering column and a junction box for the headlamp cables was incorporated into the wiring. This was a Lucas C 31 or C 30L, with a press-on tin cap. Black enamel continued to be the standard finish for headlamps on all factory-bodied cars.

A different rear lamp, a Lucas T101, was introduced during April 1932 from car number B5-6100. With it came a new type of number plate, to which it mounted by two bolts.

Between chassis numbers 148200 and 159534, the starter switch bracket was welded to the floor rather than screwed as it had been previously and, although not identical to its predecessor, it still carried the same ST8 part number. Immediately after this an updated M35 starter motor was introduced, now bolted directly to the offside of the crankcase and incorporating a Lucas ST9 switch screwed to its top; this is a lever type and is operated by a cable terminating in a knob on the dashboard.

At the same time, from chassis number 159534, the dashboard was completely redesigned and a more comprehensive layout introduced, comprising a Smiths electric petrol gauge, Lucas PLC5 ignition and light switch with summer and winter charging facility, and a key. The latter would not have been proof against any but the most bungling burglar or child, as it was of the simple spade variety. The ammeter employed was now a Lucas C213/A17 and the ignition warning lamp, also by Lucas, was a WL3. To illuminate these added instruments, a chromium-plated dash lamp was provided.

For the 1933 model year the vacuum windscreen wiper was updated by fitting a Lucas 30G in place of the earlier 30F model. The ignition coil had for some time been standardised as a Lucas 6Q6.

In August, from chassis number 179368, the head and tail lamps were changed, to Lucas type LB130P and ST38A respectively. To go with the former a Lucas FS12F foot-operated dip switch was used, and a 39C switch was used to activate the brake light on the latter.

The introduction of the Ruby and associated models from chassis number 198596 saw further changes to the electrical equipment. Headlamps were changed to the LBD130 model and sidelamps appeared, for the first time on a factory-bodied car, mounted on the wings. External box trafficators had been fitted to the Seven since the previous year so it was not surprising to find the new model fitted with built-in versions, firstly type SE26 followed by SF28A, both by Lucas. Instruments were now illuminated from the rear, operated by a PS6/L switch, so in consequence the ammeter changed to a C226/L1 except on van chassis, which retained the protruding dash light. At last the old vacuum wiper was replaced by a Lucas MT7 electric unit. At the same time a more modern horn was introduced in the form of a Lucas HF722, and the latest DK4A distributor introduced, incorporating automatic ignition advance.

From chassis number 216890 the improved type 39 stop light switch was introduced and from chassis number 220395 the headlamp wiring was altered to a separate harness with each lamp. Shortly after this, in September 1935, from chassis number 230952, a CFR 2 cut-out was introduced and some of the colour coding for the cables altered. Just before Christmas, at chassis number 236210, the electrical system was changed from negative to positive earth.

During 1936 the Lucas 4J junction box replaced the previous type and some cars, usually sports

91

Original Austin Seven

The CAV rear lamp fitted to early cars, here on a 1923 AB tourer; note also the cast aluminium number plate. The red triangle was fitted by owners of cars with four-wheel brakes to proclaim the fact proudly and warn motorists with mere two-wheel braking to beware of their superior stopping capabilities.

The Lucas T101 as used on the Ruby, this example being a 1937 ARR. The other rear lamps are not original, but the problem of how safely to light the rear of a car such as this is not an easy one to solve in a sympathetic manner. In this picture the budget lock key is inserted ready to open the spare wheel cover; the two slots in the base are for the luggage rack to protrude through when it is folded out.

The Lucas T101 or 'pork pie' rear lamp – here on a 1934 RP saloon – mounts directly onto the rear number plate, which has an extended end for this purpose.

The Lucas trafficator: not very visible in the twilight amid today's traffic, but the latest thing during the early 1930s.

Rear lamps on a 1934 Speedy. Originally there would have been only one lamp on the offside, but some owners fitted their cars up in this fashion to remain legal while abroad.

92

ELECTRICS & LAMPS

models, were fitted with a Lucas MT12 wiper motor as an alternative to the MT7. During the spring the dip switch was changed to an FS22; its introduction was haphazard among the models but all received it by chassis number 243013. In the summer, from chassis number 248007, a Lucas C35M dynamo came into use and around the same time some cars were fitted with M35G starter motors as an alternative to the M35A. Towards the end of 1936 the trafficators were updated by the introduction of the Lucas SF24, the fuse box for these remaining the same Lucas BG252 component.

At the beginning of 1937 the starter switch was replaced by a Lucas ST10 and later on in the year a DW3 wiper motor began to be fitted, eventually replacing the MT7.

For the 1938 model year the headlamps were altered to Lucas LBD 131 and the sidelights to LBD 109; the following year the headlamps were changed again, to LBD 133 units.

TOOLS

For such a small, cheap car, the Austin Seven was sent out from the factory with a surprisingly comprehensive tool kit. Over the years this was altered a little but it would be impossible within the confines of this book to list all of the numerous variations.

The following items were found in a typical tool kit, this one dating from 1927 and taken from factory parts list number 353L: double-ended spanner (⅞in × ½in), double-ended spanner (⁵⁄₁₆in × ⅜in), double-ended spanner (⁵⁄₁₆in × ¼in), sparking plug box spanner, tommy bar, carburettor jet key, magneto spanner, sparking plug and tappet clearance gauge, tappet adjustment spanner, screwdriver, tyre lever (Dunlop), adjustable spanner (4in), box spanner (⁵⁄₁₆in × ¼in), tool roll for above, box spanner (⅜in × ¼in) hub cap and steering column socket spanner, wheel brace, valve lifter, tyre pump, jack, grease gun (Autoram type), spanner for dynamo casing nut and cylinder nut, adapter and grease nipple for front and rear hub greaser, trademark cleaning plate, cylinder head joint washer and combination pliers.

The following could also be added as extras: cylinder head lifting screw, tommy bars for lifting screw, spanner for thrust and adjusting nuts, hub extractor screw, hub extractor and screw.

The double-ended spanners supplied in these kits have the word Austin on them and are worth looking for in old boxes of tools – they quite often turn up. An exact list of the tools supplied for any particular model will be found in the parts list appropriate to the year of the car; these can be found and bought quite cheaply or are available in reprint form along with the appropriate handbook.

The tool kit on many models (this is a 1926 R saloon) lived under the driver's seat in a Rexine roll. It is far more comprehensive than one might expect on an inexpensive car, but in those days motorists had to be able to sort out some maintenance problems for themselves. Template is for polishing the nickel badge on the radiator without marking the surrounding paintwork.

93

AMERICAN AUSTIN

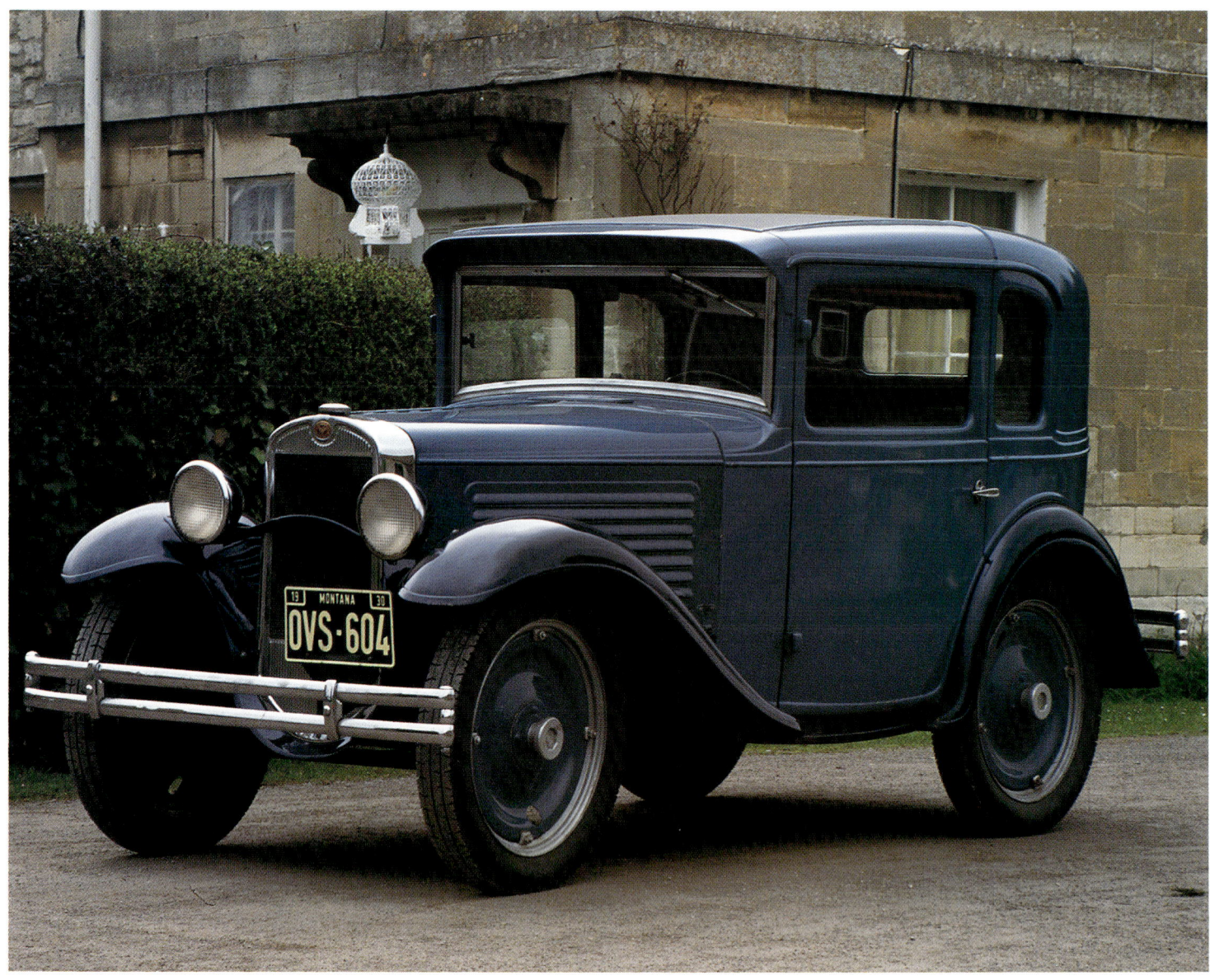

The American Austin was a cleverly scaled-down version of styles normally found on much grander vehicles. Owned by David Williams, this coupe dates from 1930 and carries chassis number 1187. The windscreen frame is not original; the old steel one had rotted out and this one was constructed from sections of a frame from a larger English Austin.

The story of the Austin Seven produced in the USA during the 1930s should perhaps borrow its title from the book by Charles Dickens – *Great Expectations*.

Sir Herbert Austin, having established his little protégé in England, had been looking abroad for other markets and had successfully negotiated licensing arrangements in both France and Germany, but so far failed in North America; a deal with General Motors was attempted that came to nought partially due to shareholders' lack of confidence. Towards the end of December 1928, Austin, accompanied by his wife, set sail for America. He had booked a stand at the forthcoming New York National Automobile Show and in the hold of the ship were four Sevens to display there. Upon his arrival he took large display advertisements in several newspapers to the effect that the Austin Motor Company was interested in hearing proposals from anyone who would be prepared to manufacture and market the Austin Seven in North America.

Two serious enquiries ensued. The first was unable to raise sufficient capital but the second, fronted by one Elias Ritt, was by the end of February 1929 incorporated in the State of Delaware as the American Austin Car Company. Finance was largely raised by two share issues which were over-subscribed, with even Austin and his company taking an option on 50,000 between them.

Ritt, along with various financiers, naturally became a director of the fledgling company. He was also vice-president of the Butler County National Bank in Pennsylvania, so it was not altogether surprising that a large disused factory in Butler was chosen to house operations. This made sense as it had been the home of the Standard Steel Car Company before it went out of business in 1923. It had also made railway carriages and produced steel,

American Austin

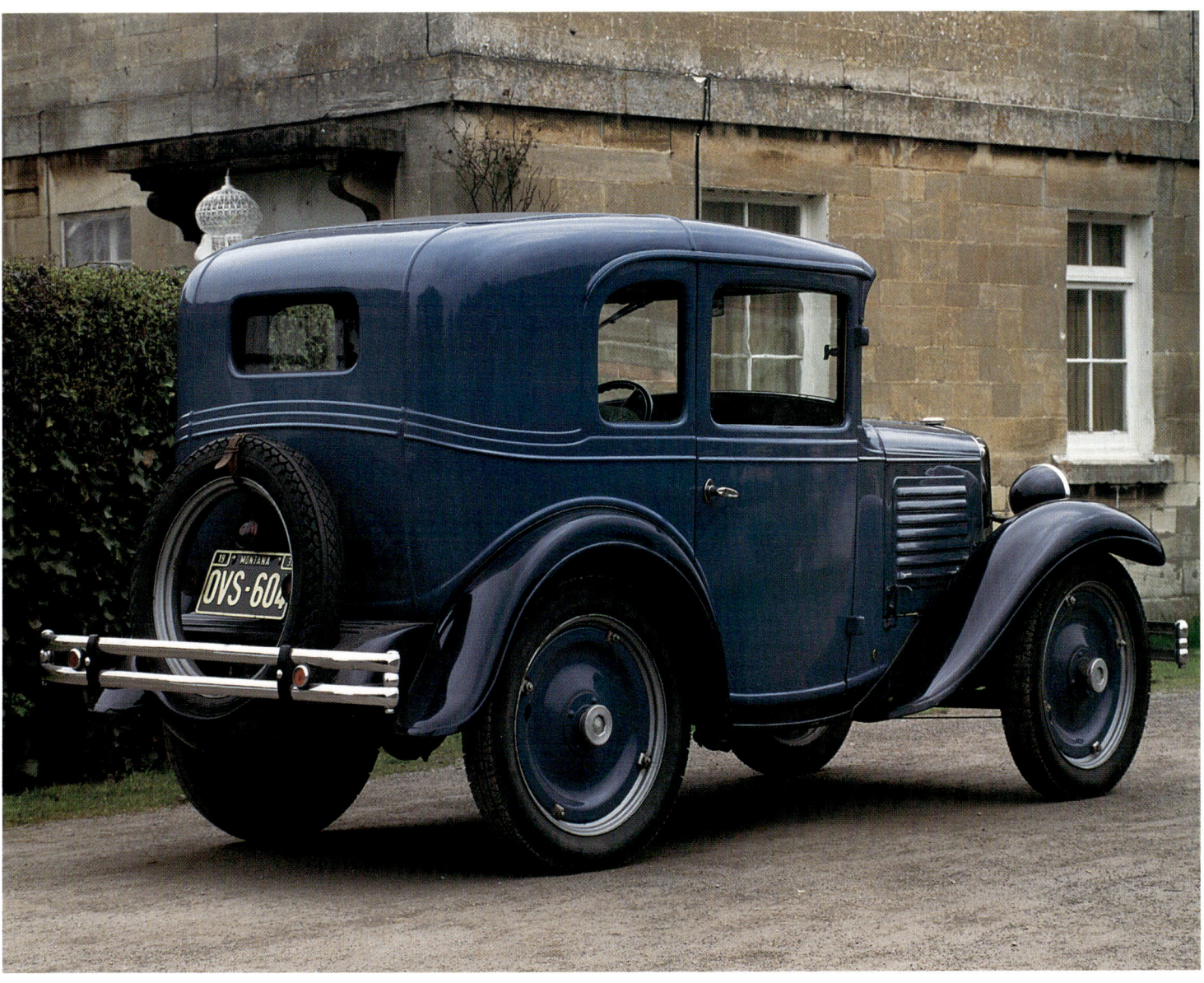

The wheels bear more than a passing resemblance to those of a Packard; rims are demountable, so the spare is a rim carrying the inflated tyre.

and much of the workforce still lived locally.

An Arthur J Brandt was elected as president, his experience in this field stretching back to 1919 when he first became an executive for General Motors. Development was carried out at Detroit due to the expertise being unavailable in Butler, and part of the agreement with Austin, other than royalty payments, was that the chassis and engine would remain essentially the same.

Stylistically the bodywork of the English versions was considered unsuitable for the American market so two firms, the Murray Corporation and the Hayes Body Company, were asked to come up with something more appealing to Americans.

Hayes won the contract, its designs executed by Alexis de Sakhnoffsky, who was already well-known for his stylish work on far more glamorous cars. Considering the parameters regarding comfort, accommodation and light weight, the results were exceptional and truly gave the impression of a large car in miniature, particularly in coupe form.

By January 1930 three prototypes had been built – a roadster, a coupe and a delivery van. Once again Sir Herbert ventured across the Atlantic and at the beginning of the month made a tour of the factory at Butler as well as attending a civic reception in his honour, during which he outlined progress made so far and announced that the transfer of operations from Detroit to the factory would begin within a month. The prototypes were taken to New York where the Hotel Shelton was used for a trade showing to coincide with the 1930 National Automobile Show. Austin was ecstatic with the response and it was put about that more than 50,000 orders had been taken in the first week and that dealers were falling over themselves to become agents.

Several months passed before production began, on a new U-shaped assembly line. Bodywork arrived from the Hayes factory fully trimmed and painted so that only final fitting was necessary at the Butler plant, at which an estimated output was planned of around 100 cars per day, employing a workforce of more than 500.

The chassis and mechanical components were largely of the same design as the original English Seven but there were several variations. The engine was a 'mirror image', in order to keep the manifolding on the opposite side to the steering column. The rear axle was a semi-floating type with the casing made up of a pair of cast iron centre sections into each of which were pressed, then welded, the axle tubes. Visually the axle was quite different from the one made by Austin, although the ratio of 5.25:1 was shared. The three-speed gearbox was built, to Austin design, by the Warner Gear Company and with the standard back axle had overall ratios of 5.25:1, 9.7:1 and 17:1. Although operated by cables in the same manner as their English counterpart, the brakes had larger 8in drums.

Various components were bought from outside suppliers, such as the carburettor from Tillotson and the electrical equipment for the most part from Auto-Lite. The rear of the chassis frame had a pair of extensions, not unlike the Ruby type, to support the body and carry the bumper. The 18in diameter disc wheels had demountable rims fitted with 3.75-section tyres.

By 3 June 1930 deliveries started to the 94 distributorships that had been granted and by 15 June the production target of 100 cars a day was attained. By the end of June dealers began to display the first cars for sale to the general public.

The curious flocked to see this little vehicle which publicity proclaimed as a desirable second car offering both economy and ease of parking. It enjoyed a brief fashionable period among the cinema colony in Hollywood and several stars took delivery of some of the first cars to be sold.

The sales department of the factory let it be known that more than 184,000 cars had been ordered and over 19,000 dealerships had been applied for. It all seemed too good to be true; and it was. Through no fault of its own other than its unfamiliar diminutive size and some unfortunate advertising copy, the American Austin became more an object of fun than anything else. Thus it became widely known for various antics, ranging from King Arthur's knights charging into battle driving roadsters in the film *A Connecticut Yankee in King Arthur's Court*, to coupes and vans advertising wares as diverse as bust enlargement preparations and peanut butter. This attracted attention in plenty but in the end the man in the street was unwilling actually to own and drive such a device – and total sales by the end of the year amounted to just over 8000 cars.

American Austin put on a brave face and pro-

Even before the American Austin became Bantam after its liquidation, the radiator badge incorporated this ferocious-looking cock bird. Wheel centres were finished with pressed aluminium screw-on cap.

Hayes Body Corporation produced bodies for all early cars. An exquisite little badge – hardly echoing the company's origins! – was mounted low down on the right-hand side of the scuttle.

ceeded to make the same error made by motor manufacturers before and since: to boost sales of an already slow-selling product it enlarged the range. Added were a Business Coupe and a Cabriolet. The former had but one seat with the rest of the space given over for a salesman's wares, while the latter was in fact a Business Coupe with full seating, a padded top and hood irons; in no sense was it a cabriolet but Ford had already used the more appropriate title of Sport Coupe for its Model A of the same style. The original coupe was now available in both standard

American Austin

Manifold castings look similar to English versions but are reversed; the whole engine is a 'mirror image'. Carburettor is a non-original replacement. The Autolite distributor sits up high; headlamp glasses are period non-dazzle accessories but were not original equipment.

and de luxe versions and the roadster was henceforth normally referred to as the Runabout.

All this was to no avail and by the end of 1931 a mere 1279 vehicles had been sold, the majority of them roadsters. The American Austin factory closed early in 1932 with 1500 unfinished cars lying at the factory and debts of more than $1 million.

This would have been the end had it not been for one man, Roy S Evans. A self-made millionaire at a very early age, among his various business enterprises he had specialised during the depression in volume buying of surplus new cars at knock-down prices, the benefits of which he passed on to the customer. Very soon he had become probably the largest car dealer in the southern states. And so when he heard that the entire stock of American Austin cars could be had for a good price, Evans, aware that although not a sales success they were well-known, took the lot.

In the deal was a proviso that the cars would be finished and once this was done Evans had them transported to Florida. Priced at $295, as against the original $445, they sold out easily. This success spurred on Evans to negotiate with the factory to produce further cars, which it agreed to do providing the financial responsibility was his. At the same time he purchased surplus stocks of parts suitable for the Austin from the manufacturers and suppliers at very attractive prices and also was instrumental in the transfer of body manufacture from Hayes to the Butler factory – this would save time and money.

Literally flying around the country (he owned and flew his own plane which he put to good effect when setting up his various dealer networks), Evans in no time at all put together a chain of 80 dealers to sell the Austin. Things improved, a little, and by the end of the year nearly 4000 cars had been delivered.

Towards the end of 1932 various improvements were made for the following season. Bodywork was given a minor facelift with a vertically louvred bonnet, and the coupe windscreen now had a gentle slope; in addition screens were higher on this and the roadster. Radiator shells were chromium-plated and there was a custom line introduced, with special colours and Bantam radiator cap among other gimmicks.

Of more interest to the technically minded were the strengthened engine mounting lugs now incorporating rubber bushes and a greater oil supply to the timing gears, the latter achieved in part by the elimination of one of the oil return holes. The steering box and rods had their wall thickness increased for greater strength and the ratio altered from just over 6:1 to 9.5:1. At the same time the track grew to 42½in. Other alterations included the provision of a filter in the petrol filler, a zigzag-pattern radiator core which gave a larger cooling surface, and a larger horn. Headlamps were now mounted on the wings rather than on a tie bar, and single-bar bumpers appeared, both of these changes occurring in a rather haphazard fashion as stocks were used up.

In the spring of 1933 the gear-driven dynamo was abandoned and from then on both this and the fan were driven by a single V belt; the casting on top of the timing case was correspondingly altered to accept a distributor. Small alterations were made to pistons, clutch release bearing and camshaft lobes around the same time. Some of these changes were primarily to improve mechanical silence, and to assist in this direction dashboard and floors were insulated with felt.

97

Sales in 1933 approached 5000 and, although not the runaway success that had been expected, the American Austin was surviving. In addition, interest was shown by the American forces, which had heard of the British Army using specially-built Sevens. One was ordered with oversize tyres and skimpy bodywork, and some extremely low-profile prototypes were built. Sales of passenger cars began to fall off in the beginning of 1934 and by June the firm filed for bankruptcy, but production was kept going until the end of the year, around 1300 cars being completed. Even during this period, Evans, still enthusiastic, brought out another model. This was the 475 and was in truth very little different apart from sloping bonnet louvres and safety glass.

The factory stopped production at the end of the year but kept on a much reduced workforce for servicing and factory maintenance. A 575 series was listed in some trade publications for 1935 but this in fact referred to a handful of new old stock cars in storage; these were sold and were not a new model. The excellent machine shop within the factory also undertook outside work when required.

Evans refused to accept that it was all over and with his friend and legal representative William Ward began negotiations to acquire the factory and all that went with it. The pair persuaded the Pullman Standard Car Company, which held a mortgage on the property of $150,000, and Butler County, which was owed some $35,000 in back taxes, to reduce the debt and interest payable in the eventuality of Evans buying the company. Although a wealthy man, he was not prepared to buy the concern for anything like its valuation of some $10 million: the Federal court, with the agreement of the majority of the creditors, at length sanctioned its purchase by Evans for a token payment of $5000.

Brokers on Wall Street were appointed to raise $2 million capital and shares were issued with the brokers' commission payable in shares. The new company was registered as The American Bantam Car Company on 2 June 1936, Evans taking the name from the motif used by American Austin.

An impressive brochure was printed which, apart from illustrating the proposed new models, introduced the major players who were now behind the company. They were as follows: Roy Evans as president; Martin Tow, an Argentinian who for many years had owned a large department store in Buenos Aires, as chairman and treasurer; Harry Miller, the brilliant racing car designer and builder, in charge of engineering; Thomas Hibbard, one of America's foremost coachwork specialists, as designing engineer; Peter Beasley, president of the Detroit Aircraft Corporation, as secretary; and R H Blackie, an Englishman who had previously worked with Vickers in his home country and latterly with American Austin, as production manager.

It is obvious that some of these men would have had little to do with the Bantam, but had been given seats on the board in return for lending their names to the company in order to make sure the required capital was raised. Indeed one of the new models shown in the publication was a Custom Roadster supposed to be styled by Hibbard and 'engineered for speed and stamina by Mr Miller. Will be available supercharged'. Miller, it is true to say, did work at the factory for a short while and did look at the manifolding on the Bantam engine, but he spent most of his time there playing with a four-wheel drive vehicle he constructed and completing a racing boat engine. Hibbard, as far as is known, had even less to do with the company.

The ill-fortune that had seemingly followed the first company struck again when it was learned that the stockbrokers who were supposed to be selling shares had merely sold those assigned to them and made off with the proceeds. Evans, not to be put

Two views of the two-seater that followed on the heels of the coupe when American Austin production began in 1930. With weather equipment erected, the design of the sidescreens, complete with flap for the driver to make hand signals, can be seen, as well as the oval rear window and contrasting piping. In the other view, showing Buster Keaton and his two sons, pinstriping is visible on rear of body and wheels. Did he own this car or was this merely a publicity shot, one wonders?

Revised styling – including radiator cowling, bonnet, wings and rear valance – accompanied the American Austin's rebirth as the Bantam 60 in 1938. Roy S. Evans, rescuer of the foundering project in 1932, is in the passenger seat, beside W.H. Cameron of Car Production Ltd, Australia – the shipment of 50 chassis to Australia is being celebrated. Standing are former manager and vice-president R.O. Gill and his successor F.H. Fenn.

The truly frightful Boulevard Delivery version of the Bantam was based on a roadster with the rear end converted into a load area and decorated with opera lamps.

down, arranged a loan of $25,000 from the Reconstruction Finance Corporation against the factory.

A development period began which was really just a rehash of the existing formula with a few modifications. Alexis de Sakhnoffsky was persuaded to modernise the looks, which he did with a new radiator cowling, bonnet, wings and rear valance. For this he was paid a paltry $300, the bargain increased by the low tooling costs of $7000 for all this. To avoid being liable for any more royalty payments to Austin, although Sir Herbert had long since lost interest in the flounderings of the Butler concern, it was deemed necessary to change the engine a little. Out went the ball and roller bearing crankshaft bearings in exchange for plain white metal ones, these having the added attraction of lower cost. The chassis frame was strengthened and half-elliptic springs now supported the rear axle, which was a bought-in and modified unit. A different clutch with greater pedal movement was now employed, and a three-speed gearbox by Warner (and originally used by Studebaker) coupled to an open propeller shaft completed the car's notably robust transmission.

Braking was improved with revised geometry, although the American factory for whatever reason – probably one of economy – followed the example of the Longbridge works in England by refusing to adopt hydraulic brakes. Steering was now cam and lever and the new design of 15in road wheels were shod with either 5.00 or 4.00 section tyres depending on the model: 16in wheels were a catalogued option for vehicles intended to be used in areas with rougher terrain.

The new car's eventual launch was assisted by nearly $1 million of previously unsold shares being marketed by an enterprising broker and at last, for 1938, the new model went on sale as the Bantam 60, the numerals at least suggesting the top speed if not promising it.

To be honest, it was easy to see why de Sakhnoffsky either asked for or was happy to accept a token payment for the restyling: the lines of their forebears had been ruined by the addition of a lorry-like radiator cowl and modern teardrop wings while retaining the early 1930s basic body shape. The Bantam really did look most unfortunate; and the roadster surely had become the forerunner of Noddy's little car, designed, if not made, in Toytown.

Two more models were announced during 1938 – the Foursome and the Boulevard Delivery. The latter was a travesty of a thing, a roadster with the rear removed and replaced by a large box garnished with opera lamps, the whole normally painted black. This Draculan creation was intended to be used as a delivery vehicle by smart shops and indeed a number were sold, but sales of all models for the whole year amounted to only around 2000 cars.

As there were purported to be more than 400 dealers in the US and 46 other countries, this was not a particularly good effort unless the car's size and new styling really were insurmountable obstacles – after all people had come to expect American cars to be large and old habits are hard to break. Those who actually owned the cars, on the other hand, were very enthusiastic, especially regarding their tenacious roadholding, but it was to be a good few years before the American buying public cared much about that kind of thing.

Right at the end of the year yet another model became available, a station wagon with wooden body made by the Mifflinburg Body Company. Chassis intended to carry this bodywork had extensions welded to the rear at the factory before leaving for the coachbuilder.

For 1939 prices were reduced and ranged from $339 for the standard coupe to $565 for the station wagon. The list of models became ever more complex; for instance a standard roadster at $449 for college students or a de luxe roadster complete with a host of goodies such as whitewall tyres, trim-matched horn button and hand-tooled stainless steel dashboard.

In February 1939 a small export success was achieved and 50 chassis were shipped to Australia where they would receive local bodywork, and Roy Evans, when he learned that the Austin Seven had been phased out in England, seriously began to think of moving into Europe with assembly plants in Britain and Belgium.

During the summer yet another new model appeared, the Hollywood, which was another clever facelift of a coupe, with a soft top and customised paint and fitments.

In the autumn the engine was enlarged to just over 800cc and given three main bearings. The ignition had for some while had automatic advance and the carburettor was by Zenith. This engine was named the Hillmaster and went into the Super 4 model of 1940. The brakes also came in for further attention; feedback through the pedal due to slightly eccentric stamped drums had long been a source of annoyance, so the shoes were allowed to move slightly at their pivots. This 'floating shoe' cured the braking problem but not the more important one of faltering sales. The year closed with just over 1200 cars sold.

For the new year, 1940, hydraulic shock absorbers by Monroe appeared, so-called 'variable ratio parking master steering' was trumpeted, and the headlamps were repositioned – a rather unhappy move – further out on the front wings. The list of accessories now included a futuristic radio for the dashboard. The Speedster had been discontinued as well as the unfortunate Boulevard Delivery, and in their place came the convertible Coupe and convertible Sedan. For this model year the pair of horizontal chrome side flashes on the sides of the rear wings on convertibles were replaced by four slanting ones.

All this and more was offered in the enthusiastic publicity material, but by the summer the company was without operating capital and sales for the whole of 1940 were around 800 vehicles. The following year, 1941, really was the end of private cars as the last 140 or so were sold off as current models.

But this was not the end of the company. During the early part of 1940, at the same time as the passenger car side was foundering, the company had been building an experimental four-wheel drive vehicle at the behest of the War Department. After undergoing stringent testing at army proving grounds, the prototype, the true antecedent of the famous Jeep, proved ideal for the job and a further 70 were ordered and in production before Ford and Willys versions became a reality and took over. These two companies had the greater production facilities necessary to fulfil the huge demand as the US was entering the war. Some 2500 Bantam Jeeps were produced, however, and the majority were shipped to Russia for use on the Eastern Front.

The war saved the Bantam company, which went on to make thousands of trailers, torpedo motors and other arms, but car production was never resumed. Roy Evans sold his interests in the company during 1946 and the name finally ceased to exist in 1955 when it was absorbed by American Rolling Mills.

PRODUCTION FIGURES

AMERICAN AUSTIN

1930	8558
1931	1279
1932	3846
1933	4726
1934	1057
1935	140*

AMERICAN BANTAM

1938	2000*
1939	1227
1940	800*
1941	138

* These figures are from the best source available in the absence of exact information.

COLOUR SCHEMES

Limited information is available for these cars but this is a guide to some models.

AMERICAN AUSTIN COUPE
Body Everglades Blue with Gold pinstripes, Serpentine Green with Cream pinstripes, Sumatra Beige with Gold pinstripes **Wings, front and rear aprons, bonnet catches, headlamp bar and bumper irons** Black **Wheels** Body colour with galvanised rims **Radiator shell** Chromium-plated with recessed portion body colour.

AMERICAN AUSTIN ROADSTER
Body Black over Cream with Cream pinstripes, Red over Cream with Cream pinstripes **Wings, front and rear apron, bonnet catches and headlamp bar** Upper body colour **Bumper brackets** Black **Wheels** Cream with upper body colour rims and pinstriping.

BANTAM, 1938-39
Body, including wings, headlamps and wheels Bantam Green metallic, Golden Beige metallic, Bantam Grey, Bantam Cream, Black **Two-tone option** Granby Red side panels with the remainder in black.

BANTAM, 1940
Body, including wings, headlamp and wheels Bantam Grey, Silver Green metallic, Sunset Maroon metallic, Black **Deluxe model additions** Vermillion pinstripe on raised bonnet and body side mouldings; Vermillion pinstripe on raised moulding on rear of roadster; single or more usually double pinstriping on wheels (Silver on Black or Maroon cars, Red on others).

Notes Leather upholstery was tan on green cars and red on others. Deluxe cars were fitted with Firestone whitewall tyres.

DIXI & BMW

During the middle of the 1920s Austin Sevens were imported into Germany by Koch & Weichsel of Berlin. It managed to sell quite a few cars, but Sir Herbert Austin could see that it should be far more profitable to sell the manufacturing rights rather than remain dependent on this firm to market his products, and also leave himself open to the vagaries of currency and import restrictions. For a while he looked for a suitable company that was also keen to produce the Austin Seven under licence, so when, in 1927, he had the opportunity to meet Jakob Schapiro he was enthusiastic to do so.

Schapiro, who was in his early 40s, had started as a penniless immigrant and had built up a very successful industrial empire specialising in investment and management. Among his many interests were various firms connected with the motor trade including NSU, and he was a major shareholder and board member of Daimler-Benz. Another company under his control, Gothaer Waggonfabrik, owned a car manufacturer named Dixi Automobil Werke AG of Eisenach. This firm, established in the 19th Century as Fahrzeugfabrik Eisenach, had begun by manufacturing a version of the French Decauville under licence. This was named the Wartburg after the ancient castle which overlooks the city of Eisenach. The name of the company had been changed to Dixi in 1904 after the founder, Heinrich Ehrhardt, had pulled out. In 1927 it boasted two models; the excellent but slow-selling 6/24 and a new 3.6-litre 'six' which at 10,000 marks was overpriced in the market conditions prevailing in Germany at that time.

Schapiro saw an easy and profitable way into the small car market, which was wide open in inflation-ridden Germany of the 1920s. Negotiations with Sir Herbert Austin were speedily concluded and Dixi's parent company, Gothaer Waggonfabrik, bought the licences which enabled it to manufacture, exclusively, the Austin Seven in Germany and Eastern Europe. A proviso was insisted on by the various financiers that German raw materials be used in the cars' construction and a minimum production of 2000 vehicles per year was also agreed upon, although the company hoped for at least 10,000.

At the end of the summer of 1927 the parent factory at Longbridge sent out to Eisenach a number of complete cars and enough components for Dixi to assemble more. One hundred cars were assembled this way, and although these were named Dixi they were in fact pure Austin Seven.

By 1 December 1927 the German factory was tooled up and started making its own version which it named the DA-1 3/15PS, which stood for Deutsche Ausführung, meaning 'German Version'; both taxable and developed horsepower were denoted by the 3/15.

Technically the whole car was almost identical to the English version apart from, in common with other 'foreign' Sevens, the engine, gearbox casing, steering box and other components necessary for the conversion to left-hand drive, being almost mirror images of their English counterparts. All nuts, bolts and studs were metric.

Electrical equipment was by Robert Bosch; coil ignition was fitted to the actual Dixi-manufactured cars but of course the very first Dixis – merely mid-1927 Austin Sevens assembled in Germany – had magneto ignition. Shock absorbers were also by Bosch. Compression ratio was a little higher than the English cars, at 5.6:1, achieved by the use of a slightly thinner cylinder head gasket, which marginally increased power output. At first a Zenith 22K carburettor was fitted but after around 2000 cars had been produced it was changed to a German-manufactured Solex.

Although there was little chance of mistaking the parentage of the early Dixi, even to the bodywork of the tourer and sedan which was almost pure Seven,

The first Dixi, the DA-1 (top), was externally indistinguishable from the English AD tourer apart from left-hand steering, headlamps, Dixi badge on the radiator and Centaur mascot. An early example of a BMW 3/15 DA-2 (above) which the factory designated a Sport model. Scuttle design is reminiscent of the Model A Ford, with a fairly narrow bonnet and the scuttle gaining width just in front of the doors. These two-seaters were produced between 1929-31.

the Eisenach factory did make small changes to render the product more palatable to Teutonic taste. For example, the steering wheel was given a polished wooden rim.

In common with the English Seven, chassis were available to outside coachbuilders and consequently a small number of cars were produced with German specialist coachwork by such firms as Buhne, Buschel, Ihle and Sindelfingen during the life of the car in both Dixi and BMW guises.

Four standard cars were listed; tourer, roadster, coupé and sedan, the German company opting to use the American bodywork terminology to describe the models. The tourer proved the most popular and made up the bulk of production. It and the sedan were based on the English AD tourer and R saloon respectively.

When the Dixi DA-1 appeared in Germany it had two main rivals, the Opel 4/12 and the Hanomag 2/10, both fairly ghastly contrivances that in any case had only two seats. So it was not surprising that the Dixi overshadowed them, and by 1929 they had disappeared. Opel abandoned manufacture of that class of car altogether while Hanomag tried a bit of mild plagiarism with its 3/16, which was visually similar to the Dixi. A fresh competitor arrived in 1928, however, in the form of the twin-cylinder, two-stroke, P-type DKW, and this would outlive the German Seven.

By the end of 1928 just over 6000 Dixi DA1s had been sold, but before that time events had taken another turn.

For some time the firm Bayerische Motoren Werke, based in Munich, had been toying with the idea of going into car production. BMW had itself been born out of Rapp Motoren Werke which had been founded in 1913 by Karl Frederick Rapp to build and supply engines mainly for marine and industrial use. This company had received an enormous boost in the early part of World War I when Austro-Daimler had been unable to meet government contracts to build its V12 aircraft engine and had sub-contracted Rapp to handle the shortfall. The man who arranged this was a 29-year-old Austrian named Franz Josef Popp, a young engineer with influence and friends both in his own world and that of finance.

With this huge increase in business the company was renamed the Bayerische Motoren Werke GmbH during the spring of 1916. It was woefully undercapitalised and to try to rectify this situation Popp asked a banker friend, Camillo Castiglioni, if he could assist with some finance. This he duly did and both he and Popp became stockholders. Shortly afterwards Rapp resigned his position, further large amounts of money were injected and then in August 1918 the firm became BMW AG, a public company, with Popp as its managing director.

The end of World War I saw BMW with a large new factory building, more than 3000 employees and not very much to do. A brief respite was granted by a large contract to manufacture air brakes for railways, but this was short-lived and the firm carried on in a reduced capacity building and developing aero engines, all the time hampered by controls put upon Germany by the Treaty of Versailles.

The company began to produce lorry engines and a 500cc air-cooled flat-twin intended for light industrial use, but various German firms found that it made rather a good motorcycle engine. One of these, a local firm called Bayerische Flugzeug, had itself been an aero engine builder but like BMW had been left high and dry after the war and had turned to making steel furniture and motorcycles.

Popp had for some time seen that the way forward for BMW was to enter the car and motorcycle field and to this end, in 1922, he called upon further finance from his friend Castiglioni and bought Bayerische Flugzeug. Upon close scrutiny the Helios motorcycle it produced, using the BMW motor, was an abominable affair, and although Popp's engineers managed to improve it they decided to start with a clean sheet of paper; the result was BMW's first motorcycle, the R32. This extremely sophisticated machine came on the market towards the end of 1923 and spawned a range of BMW motorcycles whose heritage descends to the present day with their distinctive flat-twin engines and shaft drive. Always expensive and beautifully engineered, these BMWs have made superlative touring motorcycles but at the same time enjoyed great success as competition machines.

Two years after its motorcycles were introduced, BMW once again started to produce aircraft engines; these began to find successful application in the field of civil aviation and also powered several record-breaking planes.

The Wartburg DA-3 used in the Brandenburg Trial of 1930 driven by Paul Koeppen. The factory was economical with the chrome plating, even the radiator stone guard and windscreen frame being enamelled black. Upholstery of the bench seat is loose buttoned.

Popp still nurtured his desire to become a car manufacturer, however, and had even tried to persuade his financiers to allow him to buy the licence to build the SHW car designed by Professor Kamm. This advanced vehicle with a unitary body and chassis in aluminium, front-wheel drive, independent suspension and even an automatic gearbox rather horrified the powers-that-be and Popp's enthusiasm was reined in with the request that he should find something a little more conventional should he wish to press ahead with adding car manufacture to BMW's agenda.

It was thus, in 1928, that BMW, seeing that Dixi was making a success of the DA-1 and that this could provide an ideal *entrée* into car manufacture, entered into negotiation with Schapiro to acquire Dixi. Although Dixi had started well, it did have large debts with various banks, so Schapiro was quite happy to do a deal. Dixi became a subsidiary of BMW in October 1928, its debts were cleared and Schapiro received some cash and 800,000 BMW shares. For a while the existing models continued to be produced as BMW-Dixis while the new owners tested the water and at the same time put in hand some improvements.

During the summer of 1929 the BMW 3/15 DA-2 was introduced – the name Dixi was deleted henceforth. In fact very little had been done to the mechanical components of the car apart from lowering the back axle ratio to 5.35:1 and, anticipating Longbridge by nearly a year, coupling the front and rear brakes. The radiator was restyled, complete with new badging, and the bonnet, whatever the body style, had three rows of horizontal slats. Someone in the styling department obviously had a sneaking love of Model A Fords as the exact form of the previous year's scuttle had been incorporated in all body designs. This was not without good reason, in fact, as it had been noticed that increased body width would be desirable and the Ford treatment allowed this to take place without drastic tapering of the bonnet and scuttle. The front wings and one or two other aspects of the body looked as though they owed more to Dearborn than Longbridge too, although there was also influence from Rosengart's contemporary offerings across the border in France. The saloon, especially if fitted with the optional wheel discs, could easily be mistaken for the French variant from afar.

At first the cars appeared without running boards but later on it was decided to fit them, which necessitated different front wings with a modified trailing edge. The saloon had a tiny and almost useless boot incorporated into the bodywork at the rear, and all factory bodies had a double moulding along the waistline, usually infilled with a contrasting colour.

Standard body styles were the sports two-seater, cabriolet two-seater, phaeton, coupé, cabriolet coupé and sedan. The sedan was often referred to as the limousine in sales literature and was available after a while with a roll-back sunshine roof. There was also an option of fabric covering for the bodies. A delivery van was added to this range for good measure, and chassis could still be supplied to outside coachbuilders if required – but the numbers that received specialist bodywork were miniscule.

Wheel centres now had 'BMW' impressed upon them and the instruments also carried the company logo. Tyres grew to 4.00-section, trafficators were added at the base of the windscreen and the steering wheel was changed to a rather more tasteful item with the rim covered in black celluloid.

Owners were beginning to find that these funny little cars had competition possibilities and most weekends saw them taking part in rallies and careering around race tracks with varying degrees of success. Encouraged by this, the factory entered a team of two-seaters for the 1929 Alpine Rally and surprised itself by coming home with one of the main team prizes.

In 1930 the factory brought out its equivalent of the Ulster, bearing the model designation D-3 and named the Wartburg, reviving the old Dixi name which had local connotations. Like the English equivalent, the car was lowered by using a dropped front axle and flattened springs. The engine was mildly tuned by increasing the compression ratio to 5.8:1, using an even thinner cylinder head gasket, and upon special request it was possible to have a special cylinder head giving a ratio of nearly 7:1; the pre-war practice of adding a good proportion of benzole to the fuel was suggested with either of these. The exhaust manifold was a fabricated two-branch affair with a separate tubular copper inlet manifold, but the engine was still fitted with the same 26m Solex carburettor as standard cars, albeit with increased jets. Maximum engine speed was increased to 3500rpm and power output grew to nearly 18bhp, allowing an increase in rear axle ratio to 4.9:1.

The Light Car and Cyclecar magazine in August 1930 described the Wartburg as having the following: 'The equipment comprises electrically operated direction indicators, folding windscreen, electric windscreen wiper, Bosch shock absorbers, two-way petrol tap, and a clock, with luminous hands and figures, mounted on the steering wheel. A very convenient feature is a signalling ring, mounted on the spokes of the steering wheel, which enables the driver to sound the horn, dim and brighten the headlights or operate the direction indicators without lifting his hands off the wheel.'

Whether the luminous clock was fitted to just a few cars is not clear but the signalling ring must rate as a first with so many functions.

The bodywork was similar to that of the Ulster

Original Austin Seven

but with a more sharply pointed tail and the spare wheel mounted on the left-hand side of the scuttle rather than in the rear. Full wings, running boards, lights and soft top were supplied as standard.

Although the pretty little BMW Wartburg proved popular as a dual-purpose racer and sports car, only 150 were made. But in spite of never achieving international racing success and lacking the development of its cousin, the Ulster, it was a very important car since it started BMW on the road as a manufacturer of some of the world's finest sporting cars.

Bayerische Motoren Werke only bought Dixi as a means to an end – to get into car manufacturing relatively cheaply and quickly with minimal risk during the difficult economic conditions prevailing in the late 1920s. Popp, at least, had higher aspirations than simply building someone else's design under licence, and the company was still obliged to pay a 2 per cent royalty on all cars – half to the Austin Motor Company, half to Sir Herbert Austin.

With both these factors in mind, BMW began to design itself away from Austin Sevens during the spring of 1931. In July, the new BMW DA-4 model became available.

Off-white, the German racing colour, was a natural choice for the little Wartburg DA-3. Although it is not dissimilar to the Ulster, it would be difficult to confuse the two.

BMW had already been using on its motorcycles the motif that originated with aeroplane manufacture, so it was natural to use this on cars too; three laurel leaves were added for the radiator shell.

DIXI & BMW

This 1931 BMW 3/15 DA-4 fabric saloon has lived in England for some years and is owned by Graham Horder. The sidelights have been added since the car was in this country and are incorrect.

Although of cloth, the interior is rather more spartan than that of its English counterpart, and the dashboard is rather less informative; wood-rim steering wheel, coupled with the familiar Austin-like quadrant controls in its centre, strikes rather an incongruous note.

Technically the DA-4 was still very much Austin but for one exception – independent front suspension. Unlike most of BMW's later work, this really was the most ghastly affair. Retaining a transverse spring, with three leaves, the eyes were deleted from the ends and stub axles carrying king pin assemblies were bolted direct to the extremities, the whole anchored by radius arms and controlled by a divided track rod. For some reason it was thought unnecessary to fit shock absorbers and with the absence of these and, more importantly, any form of top suspension link, the extraordinary angles which the front wheels achieved and the resultant effects on roadholding when pressed can be imagined. The only visible alteration, apart from bodywork, was to reduce the wheel diameter to 18in while retaining 4.00-section tyres.

The choice of models remained the same but there were some alterations in the styling. The saloon came to look more like its English relation. Gone was the vestigial boot and the spare wheel now bolted directly onto a boss at the rear. The doors still had an elliptical cut-out to clear the rear wheel arches. Running boards were retained, still covered in pressed aluminium matting, as they had been on previous models when fitted. The twin waistline mouldings were replaced by a single wide bead which tapered towards the rear and was normally finished in a different shade to the bodywork. Interior trim was most often fabric on closed cars but leatherette was an option, fitted to the vast majority of open versions.

Sales had peaked in the year of BMW's takeover at over 8000, but now began to seriously recede. The Wartburg was proving hard to sell, although there was never any intention to produce it in vast quantities. In addition, it did not have the new front suspension, so it was out of line with the rest of the range and was quietly dropped.

Popp and his fellow directors were unable to decide whether the Depression or the now ageing design of their product was responsible for hindering

its sales. And its lack of performance, thanks to a tiny engine and the increasing weight of 1930s coachwork, made them think again. BMW was safe thanks to flourishing aero engine and motorcycle departments so an almost complete break from the Austin Seven design was decided upon.

The new car, the AM-1, used a backbone chassis, retaining the frightful independent front suspension and introducing swing axle independent rear suspension with twin transverse leaf springs. Hydraulic shock absorbers were now thought prudent. The engine was enlarged to 788cc by lengthening the stroke to 80mm and a pushrod overhead valve head was designed. Chain drive to the camshaft and cooling aided by water pump were two more alterations which gave the new engine 20bhp. Popp must have wondered why they had bothered.

Even more bulky bodywork for the range of saloons and tourers, coupled with pressed steel wheels, ensured that they both looked stolid and behaved accordingly in spite of an advertised top speed of 100kph (62mph).

AM-1 stood for Automobile München, and signified the break away from the Seven and the attendant royalty payments. But until BMW produced its first six-cylinder car, the 303, during 1933, it was still just possible to spot Seven ancestry in some of the AM-1's components.

The DA-4 saloon was also made as a metal-panelled version and it was with just such a car that, on 25 September 1931, BMW celebrated the production of 25,000 cars – only an admissible claim if Dixis were included.

PRODUCTION FIGURES

1927	Austin Seven*	100
	Dixi DA-1	42
1928	Dixi DA-1	6120
1929	BMW Dixi DA-1	3146
	BMW DA-2	5350
1930	BMW DA-2	6702
	BMW DA-3	90
1931	BMW DA-2	646
	BMW DA-3	60
	BMW DA-4	2621
1932	BMW DA-4	480

* Cars imported whole or built up in Germany

ROSENGART

Born during January 1881, the young Lucien Rosengart, who was later to give his name to the French derivative of the Austin Seven, started his vocation before he was in his teens. His father owned a precision engineering business and it was there that he took his apprenticeship. By the age of 15 he was taking an active role in the running of the factory.

Lucien's work in this field brought him into contact with the motor car and at the age of 19 he took out a driving licence. Shortly after this his three years' national service came due and this, served mainly in Africa, took precedence over his other interests for the time being.

Upon his return from army life, rather than carrying on working with his father, he decided to strike out on his own and to this end he rented a small workshop in Rue Saint-Sebastian, Paris. There, in company with a friend and an apprentice who had been dismissed from his father's employ, he began to produce nuts, bolts and washers, especially those required by the motor industry. This venture met with immediate success and within six months he was on the move to a larger premises in Rue d'Atlas, and at the same time increased his workforce to 10. Both cycle and motor trades were in their ascendance at that time and, due to his willingness to meet clients' needs, by 1909 he had some 60 men working for him. Quite soon an even larger factory was needed and so in 1912 one was built on the Rue de Saint-Mande which, as well as continuing with established products, began to produce dynamos. It was no coincidence that Delage that year introduced electric lighting using Rosengart dynamos.

Although Rosengart was an established and wealthy businessman by the start of World War I, he did not escape call-up and for a short while he became chauffeur to General Goupillaud. Before long it was realised that he would be of far more use in charge of his factory, so he returned and immediately put his mind to benefiting both himself and his country during the conflict. Reliability in the fuses fitted in artillery shells at that period was somewhat lacking and Rosengart came up with an improved version. The enormous artillery barrages that accompanied the trench warfare which dominated the war in Europe provided a seemingly insatiable demand for his products and once again his factory proved too small. A larger one was quickly built, followed by an even larger one at St Brieuc in Brittany during 1916.

By the end of the war Rosengart employed more than 4500 people in his various factories, which then had to revert to the industrial needs of peacetime. In addition to the products of pre-war days, Rosengart expanded the range of electrical equipment produced by his company.

Immediately hostilities had ceased that brilliant but sometimes unlucky industrialist André Citroën, who had started his gear-cutting factory in 1913 and subsequently profited from the advent of the war, as well as branching out into numerous other facets of engineering in the process, decided to enter the automobile business. He elected to produce one model, in large quantities, no doubt influenced by his knowledge of American mass-production techniques and the success enjoyed by Henry Ford with his Model T.

Citroën's car, the Type A, was only offered as a tourer and he employed the designer Jules Salomon to assist with this project. Almost as soon as it was available to the public Citroën ran into financial trouble with the old bugbear of too long a lapse between delivery and payment, exacerbated by conditions prevailing after the upheaval of the war.

The banks, rapidly losing patience, were about to foreclose on Citroën when Lucien Rosengart, now extremely rich, was approached. He had become friendly with Citroën during the war years when they had collaborated on several projects and, having faith in Citroën's overture into car manufacturing, he decided to help.

Rosengart enlisted the backing of various friends and financial houses and, in conjunction with his own immense financial resources, the banks were quickly appeased. The sole condition of this assistance was that Rosengart would supervise the factory at Quai de Javel. Under his guidance the half-track Kegresse came to fruition, and he played no small part in the birth of the 5CV. By 1922 Citroën was running well and Rosengart decided to concentrate solely on his own businesses. Shortly he would be instrumental in saving another of the great French automobile manufacturers from oblivion.

Peugeot, arguably one of the finest French car manufacturers before World War I, had got itself into a financial pickle and had a serious deficit in 1923. Robert Peugeot, having seen what Lucien Rosengart had done for Citroën, called upon his help. Once again Rosengart got together with some

Although it possesses certain charm, this 1928 LR 2 coupé is a rather less fluid design than the coupé type B made around the same time by Austin. The wheel discs, very often fitted to Rosengarts, also contrive to give the car a different look. Flashing indicators mounted below the headlamps, of course, are not original.

friends and finance houses and offered the necessary assistance provided Peugeot restructured the company, built a new factory and not unnaturally gave him a directorship. That these measures proved successful is obvious from the fact that Peugeot remains one of the very few survivors from the early years of the motor age.

At the same time as becoming involved with other companies, Rosengart made sure that his own did not fall by the wayside and during the 1920s continued to evolve new products as and when the occasion demanded. These included items as diverse as an auxiliary motor for bicycles and a novel little torch named 'Dynapoche' powered by an integral dynamo operated by squeezing the hand. Are any of these still giving faithful service in some remote French farmhouse?

The mainstay of the company remained the manufacture of nuts, bolts and associated products which by now were used by the majority of French car makers. Rosengart's work as a saviour for the floundering giants of the French motor industry was not therefore without its own benefits!

Rosengart, forever restless and seeking fresh opportunities, had toyed with the idea of manufacturing cars on his own account. His involvement with both Citroën and Peugeot gave him an insight into the commercial possibilities. The Peugeot was looking increasingly archaic with its fixed cylinder head and feeble performance, and Citroën had discontinued the 5CV. Rosengart had also decided to bring his involvement with the Peugeot company to a close, and so when he became aware that just across the channel in England Sir Herbert Austin was seeking foreign licencees for his smallest car and that a German company had secured the rights for its country, he was not slow to act.

Rosengart, a self-made man with more than adequate resources and experienced in the motor indus-

A 1932 LR 4 saloon owned by Ian Hodgson, with chassis number 60519 and engine number 23514. This car has never been restored and has survived in remarkably good condition, even though it was requisitioned by the occupying Germans during the war. Small boot with opening top lid is typically French.

Somehow so typically French that one can almost smell the lingering hint of garlic and Gauloises. The Rosengart's seats are rather more plush and inviting than those in its little relation made in Eisenach; there are no controls in the centre of the steering wheel.

try, was just the sort of person who appealed to Sir Herbert. An agreement was soon made between the two parties and in the same way as Austin had begun its arrangement with Dixi, a number of Seven chassis were sent over to France for assembly before the French company began manufacture proper. Like those behind the American Austin at its inception a little while later, Rosengart started talking in terms of producing over 50,000 cars a year.

To accommodate his new venture, Rosengart acquired a large disused car factory in the Route de la Revolte at Neuilly-sur-Seine in that area on the north-west outskirts of Paris traditionally associated with the automobile industry. Since just before World War I, this factory had been the home of the now defunct marque Bellanger, which had finally gone out of business in 1925. Since that time the premises had been used by Peugeot, and so Rosengart was able to acquire the property at very beneficial terms as part of his severance from that firm. He already had a business address in the fashionable Champs-Elysées at number 21 and so he created Elysées Automobiles as the main distributor for Paris and its environs. For a man with his contacts and reputation there was little difficulty in arranging agencies around France.

To comply with the terms of the licence the car was to be very similar to its English relation, but to assist with general engineering and any redesign considered necessary, he employed the ex-Citroën designer Jules Salomon, who had shown great enthusiasm for the liaison between Rosengart and Austin. One or two alterations were immediately put in hand but the first model, the LR1, still used some parts that had been supplied by the English factory. The switch to metric threads and any other machining necessary was obviously no problem for the Rosengart operation. Engine ancillaries were of French origin: the dynamo and starter were made in-house, the magneto was from Ducellier, and a Solex updraught carburettor was used.

Initially, one body style, a three-seater 'coupé-spider' which had two front seats and a third crosswise behind them, was offered, as well as a van. The bodywork was built on a wooden frame which was panelled in steel to the waistline and covered in fabric above. This method of construction was adhered to for subsequent models, including saloons. The bonnet had vertical louvres and there were no running boards at first. Wheel discs came as standard equipment.

Interior trim was almost invariably in the French idiom of fabric and the door cappings and dashboard were of walnut or similar fruitwood. Instrumentation was rather more comprehensive than its English counterpart, featuring a 100kph speedometer, matching oil pressure gauge and ammeter, a petrol gauge and combined ignition and lighting switch

Original Austin Seven

The petrol tank looks a bit of an afterthought on the LR 4 but is original. The engine's origins are obvious but some features, such as the fan assembly, were altered; plug leads are incongruously modern.

There can be no mistaking a Rosengart crankcase, and the timing gear cover is also peculiar to the French car. Inlet manifold still has provision for a vacuum wiper but the carburettor is by Solex.

with a spade key. All gauges had black faces and were made by OS, while the switch was by Ducellier. The petrol gauge, a calibrated drum type, was operated by the typical French system of a slender cable leading to it from a float and lever within the cylindrical scuttle fuel tank, the falling petrol level pulling the cable and revolving the calibrated drum within the gauge.

Headlamps were Ducellier type ST33 with built-in sidelights. The rear lamp was similar in appearance to the Lucas on contemporary English Sevens but it had a slender waist with a mounting flange of the same diameter as the lamp. The van employed a splendid little cylindrical rear lamp showing a red five-pointed star when lit. These were all painted black, as was the underbonnet Kifonet klaxon horn.

Very shortly the LR2 replaced this introductory car and gradually alterations to the basic English design appeared, such as the differential casing with an inspection cover at the rear and modified axle casings, and a steering box with a separate drop arm. Coil ignition was introduced with the resultant change in crankcase casing due to the magneto platform becoming redundant, and at the same time the oil filler was repositioned higher up on the shoulder of the crankcase, but still at the rear, with a tubular steel extension.

The starter, which had originally faced to the rear and had been housed in a separate casting, different from the English one, now was mounted directly on the crankcase, facing forward. A fresh dynamo was

called for at the same time and this carried the drive for the Ducellier distributor, which mounted directly onto the purpose-made body, unlike its English counterpart, which was mounted on an aluminium casting at its extremity.

In February 1929 at a hillclimb near Grasse in the south of France the firm of Rosengart gained some useful publicity when a driver by the name of Vinatier won the 750cc sports class, and also beat the 1500cc and 2-litre sports times. This was a bit of a cheat as the car was in fact one of the English

ROSENGART

By the mid-1930s Rosengart was beginning to alter the external look of the engine almost beyond recognition. The carburettor on this Supercinq should be a horizontal Solex rather than this SU.

Radiator badge carries the Rosengart script over a rose.

Chassis plate of 1932 LR 4 also gives a multiplicity of lubrication instructions.

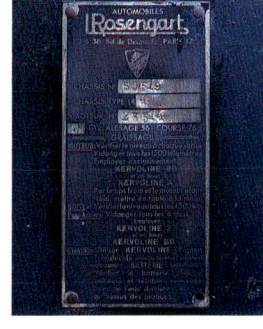

factory's Super Sports supercharged cars from 1927/28 which had been acquired for this event and dressed up with Rosengart-like stoneguard and logo. All harmless fun though, and a lot less deceitful than some other manufacturers.

The factory had by this time added further body styles, including a saloon, coupé and tourer, all with composition fabric and steel bodies, cloth interiors and discs over the wire wheels, Publicity material boasted of 2500 agents world-wide, but the vast majority were most likely merely establishments which could supply a Rosengart if called upon.

Starting on 10 August 1930, an event took place which arrested the attention of the French public and firmly established the name of Rosengart. An enthusiastic Lyon dealer, Monsieur Turel, got together with another enthusiast, Monsieur Lecot, who specialised in long-distance motoring activities. Together, with the blessing of the factory, they planned a special test for a standard Rosengart saloon. A route from Lyon via Bourg-en-Bresse to Dijon and back to Lyon was covered four times a day, a distance of some 900km. By 1 September the indefatigable Monsieur Lecot and his trusty car had covered 20,000km and then, by 4 October, the 50,000km mark was passed – the little Rosengart drinking around 55 litres of petrol and 1 litre of oil a day. At last, on 29 November, the duo's 100,000km goal was reached with the car still running perfectly.

In the autumn a new range had been announced, and was in fact the new long-chassis L2 saloon that Lecot had used for his epic drive. This came either in standard or 'luxe' forms and was in steel with fabric above the waistline. Also on the long chassis were a 'torpedo' with steel body, a van of the same material, and a pair of those French specialities of the period – one a basic steel and fabric saloon and the other a basic steel tourer, both of which could double as commercial vehicles for salesmen, farmers or tradesmen.

The long-chassis cars had a different design of chassis at the rear which allowed the fitting of half-elliptic springs in place of the quarter-elliptics used on the normal chassis, which itself had anticipated the extensions over the back axle used only on the English Seven with the advent of the Ruby.

On the normal 1900mm chassis there was the 'coupé-spider', a two-seater coupé with dickey seat. This was also available in 'luxe' and 'grande luxe' form complete with mahogany-cased vanity unit and de luxe trim. The first two versions had combination steel and fabric bodywork, whereas the 'grande luxe' bodywork was all-steel and was fitted with bumpers. Also on this chassis was the 'coupé d'affaires', a fixed-head coupé with hood irons; this again had a combination steel and fabric body. The individual front seats tipped forward to give access to the single transverse rear seat and there was a luggage trunk which carried the spare wheel at the rear. Lastly, an economy two-seater roadster, the R5, was introduced with a steel body, a bench front seat and no discs to cover the wire wheels; this was marketed as the cheapest car in Europe at 11,900 Francs.

All models had full running boards between the swept front wings and those at the rear. Silentbloc bushes were introduced in the suspension of the long-chassis cars at this time and the shock absorbers, which hitherto had been made by Rosengart, began to be supplied by Repousseau – a single type 503 at the front and a pair of 500s at the rear. The range of instruments fitted remained the same but the speedometer was now of the drum variety.

During 1930-1931 Rosengart cars competed, with a fair degree of success, in various events ranging from the Paris-Nice and Paris-St Raphael rallies to the Bol d'Or. One of the R5 variants even took part in the Grand Prix des Frontières at Chimay.

Unlike the parent factory in England, the French, under Lucien Rosengart's restive guidance, were not content merely to produce their version of the Seven and incorporate gradual modifications only when they were almost forced upon them. By 1930 the R6, a six-cylinder car of 1100cc, was introduced, using Rosengart's own sidevalve engine with a bore and stroke almost the same as the Seven at 56mm × 74.5mm. Two years later a tubular chassis was evolved for this model and a few hundred were made in this guise.

In 1931 the next development of the line appeared in the form of the LR4, which had the 2200mm wheelbase and 1050mm track of the long-chassis L2, the latter having been increased from 1020mm at the same time as the wheelbase was lengthened. Although the LR4 was still very much an Austin Seven by another name, the divergence from the original concept was accelerating. The car was becoming stouter in appearance and the bodywork had more width in order to offer the occupants greater comfort.

Various alterations to the running gear included a new rear axle which did away with the torque tube and featured a conventional propeller shaft with universal joint at each end, the standard ratio being 5.55 to 1. The noisy gears driving the dynamo were also abandoned and a belt-driven component was mounted on a large steel plate sandwiched between the block and timing cover; this cover was a completely different casting which now had a socket to accept the distributor incorporated in it. Carburation was by a Solex 26 BFVG and the electrics remained Ducellier apart from the coil, which was an SEV. Various components, although of Austin origin, were redesigned or modified on the Rosengart. An example was the aluminium water manifold on the side of the cylinder block: on the French car this was much extended and on the LR4 actually passed through a hole in the steel dynamo-cum-engine mounting plate.

The range of body styles remained similar for this improved model except that the 'torpedo' was now dropped, and modifications continued: from chassis number 34752, for instance, the capacity of the sump was enlarged from 2 to 3 litres. Comète-Mecano clutches came into use, and on cars so equipped, rather than with a Rosengart one, the gearbox casing, first motion shaft, flywheel, bearings and other associated components are not interchangeable. Three-point rubber engine mounts replaced the old four-point rigid ones inherited from the original Austin design and at the same time the carburettor was changed to a Solex 26 MVD. A while later this was changed again to a Solex 26 GHF.

In 1932 a four-cylinder 747cc version of the 1100cc car appeared. This car, the LR44, shared its chassis dimensions with its larger relation, which had the same 1050mm track as the LR4 but a longer wheelbase of 2350mm. The range of models available in these two guises during 1933 were the coupé d'affaires in drophead and fixed-head forms, the coupé-spider and a two-door saloon. The LR44 was also offered in four-door saloon or four-door commercial saloon form. The Rosengart factory often used seemingly inexplicable titles for its various models: although using the term 'Lecot' to designate the two-door saloon appears very sensible, describing the fixed-head coupé d'affaires as a 'Focof' and its cabriolet version as a 'Cofoc' defy explanation!

The LR47 supplemented the previous models in 1933 and in 1934 was augmented by the LR45, which had the same track but an even longer wheelbase of 2600mm. The range was further complicated with the addition of the LR49 during the same year. This had the same wheel plan as the LR4 but with a narrower rear track of 1118mm.

Lucien Rosengart's interests were once again straying with his introduction of a front-wheel drive car of just over 1½ litres for 1934, but he plugged on with his version of the Seven and during 1936 the

Page from 1936 Rosengart brochure shows a Supercinq in its rather more attractive earlier form. Within a year the factory, among other modifications, for some reason attached the most hideous bull-nosed radiator shell, even more ghastly than the similar styling exercise across the Atlantic by Bantam. Although the car had a bench seat and was advertised as a three-seater, the more one looks at this photograph the more the chic lady in the centre looks squeezed in by a crafty bit of montage! Rosengarts were right-hand drive.

LR4 N2 was introduced. Mechanically it was much the same as the LR4 but various changes to the specification were made. Many of these were of little consequence but one of the more obvious differences was a new cylinder head casting — except on standard models — in aluminium, with central bolt-on water manifold. At first the sparking plugs were 18mm but these were almost immediately superseded by 14mm ones. The inlet and exhaust manifolding was in one piece and a horizontal 26GHF Solex carburettor was normally fitted.

A cowled radiator grille with vertical slats, not unlike a scaled-down Delahaye 134 or 135 design, updated the frontal aspect of the car and the bonnet now had six long horizontal louvres on either side. A chromium-plated blade bumper of typical Gallic form, downswept in the centre, was also employed to harmonise the new look. Running boards were now dispensed with and more up-to-date front and rear wings were fitted. Easy-clean wheels, carrying 110 × 40 tyres on two-seaters and 120 × 40 on four-seaters, were used in conjunction with large, domed, chromium-plated hubcaps impressed with the Rosengart motif.

The new model went by the collective name of Supercinq but a large range of body styles was still available: saloons, fixed-head coupés, drophead coupés, roadsters and that charming French speciality, the commercial traveller's saloon — you could have them all from *maison* Rosengart.

A few months after the debut of the LR4 N2, the factory began to give the engine dimensions as having a bore and stroke of 55.9mm × 76.2mm. The compression ratio had been raised to a heady 6.13:1 and the maximum advertised engine speed was now 3800rpm. Shock absorbers had become standardised with Repousseau type 80 SERs all round.

For the 1938 model year the general design of the bodywork was tidied up with altered mouldings at the waistline and concealed hinges for the doors. The radiator cowl became distinctively bull-nosed and the radiator grille was split, BMW-like, into two parts and had mesh inserts to replace the slats previously used.

Production of the LR4 N2 continued until 1940 and the factory even carried on producing new models after the outbreak of war: two of these, the LR4 N2 and the LR4 R1, appeared during 1940. The latter broke new ground by having independent rear suspension and a four-speed gearbox. During 1941 a second version of the car with independent rear suspension appeared — the LR4 R1.2. The principal difference was an increase in the wheelbase to 2250mm, the previous models having remained at 2200mm since 1934.

During the occupation Lucien Rosengart left Paris and the factory fell into disrepair, so there were no new models until the end of the 1940s. Royalty payments had long since been forgotten about since the chassis of the Supercinq, introduced in 1936, hardly resembled its English forebears apart from the transverse front spring and A-frame. The engine size and retention of separate aluminium crankcase and cast iron block for its various offerings in the early 1950s still gave a tenuous link to Rosengart's beginnings in its models — 4PL, 4PL4, 4SA, 4SA1 and 4SA500.

Lucien Rosengart retired in 1953, at the age of 64, and subsequent cars were fitted with a Panhard horizontally-opposed twin-cylinder engine.

AUSTIN SEVEN DERIVATIVES MANUFACTURED BY ROSENGART

Model	Years
LR 1	1928-29
LR 2	1929-31
LR 2 G	1931-33
LR 4	1931-33
LR 44	1932
LR 47	1933-35
LR 45	1934
LR 49	1934
LR 4N2	1936-40
LR 4N2A	1940
LR 4R1	1940
LR 4R1.1	1940
LR 4R1.2	1941
LR 4PL	1950-52
LR 4PL4	1952-53
LR 4SA	1952-53
LR 4SA1	1952-53
LR 4SA500	1952-53

EXPORT VARIATIONS

Apart from chassis or complete vehicles exported to France, Germany or the USA before or during the setting up of overseas licensing and production, Austin exploited export markets in various other parts of the globe with varying degrees of success. Australia had always proved a healthy market place for the more robust makes of English car and Austins enjoyed particular success there.

Prior to World War I, the Australian government, lacking a motor industry of its own, had introduced taxes on imported cars that favoured foreign manufacturers who were prepared to supply vehicles in chassis form. These would be bodied by local firms, some of which made a very professional job and in the case of Holden went on to produce complete motor cars. Others could only manage gawky, ill-made parodies of European styles or crude utility bodies, although these were eminently suitable for the terrain prevailing in Australia.

During the first years of the Seven's manufacture complete cars were exported to Australia, but by the mid-1920s Austin succumbed to the Australian government's wishes and from then on normally shipped out cars in chassis form or bodyless from the scuttle back, apart from the floorpan. Sets of factory wings were also supplied if required.

Holden, which already constructed bodywork for the larger Austins as well as other English makes such as Morris, was quick to produce its own variation of the Seven. It quite sensibly adhered closely to the English design for a tourer – a waistline moulding and the joint between the steel rear body panelling and the Austin-manufactured aluminium scuttle were the most obvious distinguishing marks. A two-seater convertible was also produced, this having a sloping, full-width rear deck with rounded corners, the same waistline moulding and an extra moulding which followed the contours of the rear bodywork. Both models used the factory wings and running boards.

Another firm, Latrobe, produced an interpretation of the Chummy. The visual difference between this and the Holden was that it had a double waistline moulding on the doors which then converted into a single one just behind them. The join between Austin and Antipodean panelling was in the centre of the bodywork below the doors; on the Holden it was further forward. Latrobe, along with various other concerns, also produced what it perceived as a sports car on the Seven chassis, normally with pointed tails and in-house wings of either cycle or flowing form.

Towards the end of the 1920s someone, his name unknown to me, came up with a design of sports body for the Seven. This was christened the Meteor and was produced by various companies in Australia including Richardson of Sydney and Flood of Melbourne, the latter probably producing the greatest number. A pointed tail, flowing semi-cycle wings, a 'V' windscreen and a cowled radiator were features of this style, although individual manufacturers differentiated bonnet louvres and other details.

By the mid-1930s Holden was one of the few surviving firms still interested in Sevens, which had become more like customised Rubies. One of these, however, displayed a good deal of American influence, aping the popular fashion of that period as a two-seater fixed-head coupé. There was also a utility version of the open road tourer, in the French idiom, as well as a fully fitted-out version.

Another country in the Pacific to which Austins were exported in good numbers was Japan. It has been suggested in various publications that Datsun built Austin Sevens under licence in Japan, or at least had some arrangement with Sir Herbert Austin. This was not so.

Any Austin Seven imported into Japan prior to 1933 would have arrived as a complete car or as a running chassis to be bodied by a Japanese coachbuilder seeking to improve on the factory design. In order to take advantage of the Japanese taxation system, which favoured cars of less than 500cc, the first Datsun cars were designed with a motor of just 497cc. In 1933 Datsun, probably finding their cars just a trifle sluggish, increased the engine size to 750cc having first ascertained that the government were willing to raise the tax threshold to that capac-

After BMW had gone onto other things, Austin resumed export of the Seven to Germany in left-hand drive form. The 65 and Nippy models were particularly popular.

Variations of the sports-bodied Meteor version of the Seven were made by several companies in Australia, this rather fussy effort purportedly by New South Wales Motors.

EXPORT VARIATIONS

A page from the 1937 catalogue produced by the Japanese Austin dealer, Nihon Jidosha Company Ltd. This tourer has an American look about it, especially the hood, but from a few years earlier.

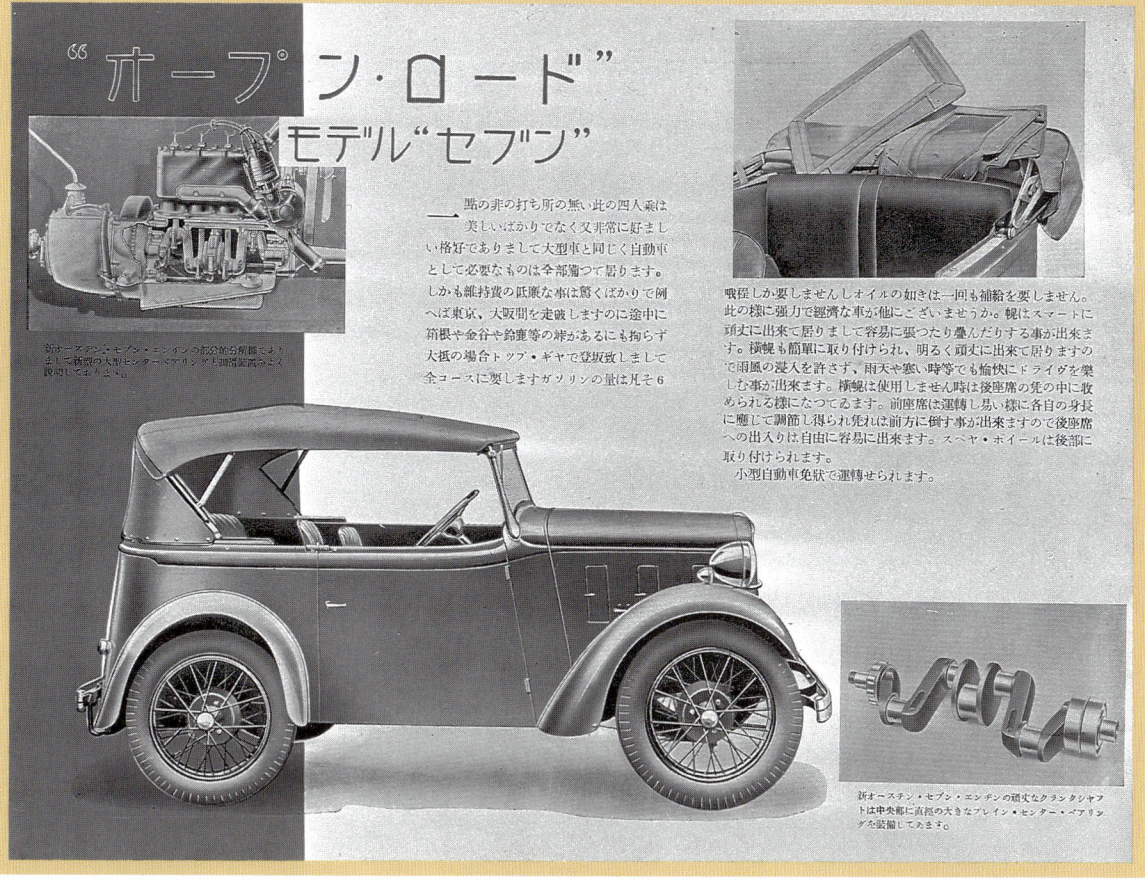

ity. Displaying true oriental cunning, at the same time they influenced the government to fix the body dimensions of a car with these taxation advantages as 1.2 metres by 2.9 metres. Thus the Austin Seven's engine then qualified, but its bodywork did not. To counter this, the Japanese importer made arrangements with Austin to receive bare chassis from England, onto which Japanese workshops would build smaller bodies which would render them eligible for this more lenient tax.

Rumours about this all reached Austin, and when he also heard that there was a possibility of Datsun importing these cars into Australia at a price lower than the Seven he instructed his agent in Australia, Crane-Williams, to acquire one of these Datsuns and send it to Longbridge for evaluation. On examination it became apparent that although it had drawn for inspiration upon the small cars then being produced by Austin, Morris and Ford, no actionable infringement of patents had taken place. Due to the regulations then in force in Japan, chassis with a wheelbase of 6ft 3in continued to be manufactured and exported to that country even after the 6ft 9in wheelbase was standardised in England. They also continued to be equipped with 18in wheels when home market cars had begun to use 16in.

The bodywork on the saloon was not dissimilar to the Ruby but the tourer had a more transatlantic look about it, with a heavy moulding along the top of the body. A delivery van was also produced. On almost all Japanese-bodied Austin Sevens of Ruby type the bonnet had three ventilator doors rather than the two fitted to home market cars, and the wings, of more scalloped form, lacked running boards.

As bodied in Japan (right), with chromium radiator and 6ft 3in chassis. Bodies were almost exclusively made by two firms, Shibaura Denki Kogyou Company and Wakita Bodies Company, although no doubt one or two other coachbuilders had a go. The same chassis length was used in conjunction with the cowled radiator (far right), which resulted in this quaint semi-Ruby 'lookalike'. The small brackets close to the wheels were for attaching anti-splash devices during bad weather; these unsightly accessories were stored in the box underneath the door.

SPECIAL COACHWORK

Louvred panel underneath the running board was a feature of the Gordon England Cup model. This one, registered on 28 May 1928, is owned by Adli Halabi and has chassis number 65179. The vacuum wiper is a Trico and the rear lights are cheap modern replacements that are quite out of place on this handsome little fabric-bodied car.

Quite apart from cars exported and bodied at their destination, there arose soon after the introduction of the Seven a market for alternative bodywork. This can be split into two categories: the first consisted of outside coachbuilders who worked in conjunction with the factory producing officially recognised bodywork that often appeared in sales literature; the second for the most part consisted of hopefuls eager to press their sometimes good but sadly often not-so-good wares upon the general public.

The first Austin Seven to be bodied by an outside firm was the 'Brooklands' Super Sports introduced in 1923. These were built by Gordon England Ltd of Putney with some of the work being undertaken by the nearby Weymann factory. Gordon England himself had been racing Austin Sevens for a while and this car, normally supplied with a tuned engine, was a logical conclusion to his activities and his relationship with the factory. Actual racing cars were fabric-covered over a wooden frame, but production versions were panelled in aluminium. The car could be had in a variety of options ranging from full windscreen, hood and wings to one stripped for racing, dependant upon customer requirements.

Gordon England's first essay into equipping Austin Sevens with bodies of his own design appealed to a very limited market but in 1925 he came up with his beautiful little Cup model. Both this and the fabric saloon which he introduced at the end of 1925 were given official blessing by the factory and could be purchased through main agents. At the same time a van could be had from the same maker. The saloon, of course, pre-dated the factory's in-house offering.

These little saloons proved so popular that Austin chose Gordon England to design and manufacture the type AD, which appeared just before the first wholly Austin-manufactured saloon, the aluminium panelled R-type that appeared in the spring of 1926.

In 1928 the Gordon England Sunshine Saloon was introduced with a roll-back fabric roof; bodies used Gordon England's own plywood floor instead of the standard Austin floorpan. Later on in the year, realising that its saloons were becoming rather dated and getting wind of the imminent announcement of the Swallow Austins, Gordon England tried to update its closed cars with the Wembley saloon, and modernised the Cup model, which became the Stadium. Neither was as successful as the previous Gordon England bodies and in any case the fashion for fabric bodies was on the wane; by 1930 Gordon England bodied its last Austin Sevens.

Another firm to receive Austin's official blessing

Special Coachwork

Like so many open cars of the period, the Gordon England Cup model does not look at its best with the hood up.

Gordon England bodies were fitted with a threshold plate and the body maker's badge below the door on the driver's side.

Gordon England Cup model dashboard with all the normal instruments plus a dashboard light and a Linkula oil gauge instead of a button. Yellow disc is one of the popular St Christopher medallions, often carrying the supplier's name, that were sometimes fitted at that time. Mounting the headlamps on brackets through the side of the radiator shell was a Gordon England hallmark. Fabric covering of the scuttle under the bonnet is typical of these cars.

117

Normally the Brooklands Super Sports (far left) was supplied in stripped racing form, but it was possible to order the car at extra cost with windscreen, hood and wings. Imagine leaving or entering the car with the hood up! A happy family (left) in their Gordon England sunshine saloon with the roof rolled back.

was Mulliner of Birmingham, using fabric construction on the Weymann principle which it already employed on the Austin 12/4 among other cars. Mulliner's first Seven was a two-seater fixed-head coupé with winding windows and an opening boot. This first appeared during 1927 and very few were made, but in time for the Motor Show a fresh design was announced: this was a fabric saloon immediately recognisable from a distance with its dummy hood irons, but also possessing several typical Mulliner features such as circular ventilators on the scuttle.

At the same time Mulliner introduced rather a splendid little fabric-covered van for use by salesmen that went by the name of the Traveller's Brougham. With its upward opening rear door containing a window, this could almost be called the ancestor of the modern hatchback. In 1928 the fabric saloon acquired a sliding sunroof and thereafter was known as the Sunshine Saloon. Due to its fashionable looks and no doubt helped by its adoption by Austin, this proved very popular and nearly 3000 were produced. In the same year a sports two-seater was introduced and, apart from its smaller doors, it resembled the factory-made metal-panelled version of the following year. In common with other Mulliner bodies on the Seven, it was built of fabric on a wooden framework erected upon a factory floorpan. Standard factory wings for the year of manufacture were used on Mulliner products but the running boards were normally covered with ribbed aluminium sheet.

In the autumn of 1929 the fixed-head coupé was modernised; the slightly sloping screen, restyled tail and roofline rendered it less gawky. A sliding sunshine roof was fitted and the side windows were now of the sliding variety, due to the wider doors now used which extended over the rear wheel arches and therefore were unable to accept a window withdrawing into them. At the same time the sports became more sporting, in looks at any rate, with a 'V' windscreen, helmet wings and a louvred metal panel running beneath the bodywork on either side. The saloon also came in for modernisation with a wider, more curvaceous body, abandoning the landau irons and introducing such embellishments as

The first type of Swallow saloon had a more bulbous radiator shell with no central division, and fewer bonnet louvres. The treatment of the false dumb iron covers and wing line sweeping up from these was also a little more rakish than on the later car. Other subtle differences in the general lines of the body make it more attractive than the model which followed.

SPECIAL COACHWORK

Pages from an Austin brochure dated 1928 show how the factory gave official blessing – and indeed publicity – to the coachbuilt offerings from Gordon England and Mulliner.

a wood-covered dashboard. All three cars aped the styling gimmick introduced by Swallow a year or so earlier, using a dumb-iron cover below the radiator, even though a car of the Seven's chassis configuration had no need of such a thing.

In the same way as Gordon England, Mulliner found that a combination of the passing fashion for fabric bodies and the economic depression of the late 1920s made it increasingly difficult to justify this side of its business. It tried a final update of both the fabric coupé and saloon, and a few metal-panelled sports models on the taller radiator chassis, but by late 1931 decided to call it a day with the Austin Seven passenger car side of its business.

Mulliner, however, was involved with the Austin Seven in another sphere as, in 1929, the firm built the rather quaint military version. In 1932 it was called upon to build more of these on the contemporary chassis, destined for use in the Near East. Part of the contract was cancelled, the surplus was sold off and some found their way onto the home market.

Probably better remembered than all these versions of the Seven, due to them being the true blood ancestors of Jaguar cars, are the vehicles bodied by Swallow. Having successfully built motorcycle sidecars in Blackpool in the early 1920s, William Lyons saw that the price differential between a motorcycle combination and a cheap car had narrowed, and this led him to expand into the car market. He could also see that if he could produce a little car with stylish looks and luxurious fittings, he would overcome the obstacle of any price difference.

To this end Lyons sensibly chose the well-proven and popular Austin Seven to test the water and, despite the debilitating effects of the general strike of a year earlier, he launched his first Swallow in the early summer of 1927. This took the form of a two-seater with a 'V' windscreen and a protuberant rounded tail, ash-framed and panelled in aluminium, even to the extent of a louvred bonnet in the same material. A bull-nosed radiator cowl was used, also in aluminium, and the very early cars had cycle wings, those at the front stayed to the brake back-plates in order that they moved with the wheels. This system is never a happy one and almost immediately there was a change to conventional wings and running boards, almost invariably aluminium-clad. An unusual and advanced option was a removable aluminium hard-top as superior protection to the standard hood during periods of inclement weather. Such other features as hardwood dashboards, opulent leather interiors and handsome torpedo sidelights were to become a hallmark of all Swallow bodies. The first two remain a feature of their descendents to the present day.

For 1928 the front of the two-seater was mildly restyled. The radiator cowling was given a smaller aperture as a result of which the starting handle protruded through the shell rather than the grille as before. The ovoid form of the base was also abandoned and the cowling terminated over a dumb-iron front valance.

Just in time for the Motor Show and shortly before moving to a larger factory in Coventry, the

Original Austin Seven

This Mk2 Swallow saloon owned by Keith Buckett was first registered on 27 November 1931 and has car number B4-1971, chassis number 138773 and engine number 139701. Apart from the wheels, the car's origins are hardly recognisable because Swallow, unlike some other manufacturers of Seven bodies, even used its own bonnet and radiator shell. Saloons constructed by Swallow, whether on an Austin chassis or another make, have this distinctive rear aspect; the rear lamps have incorrect oversize lenses.

SPECIAL COACHWORK

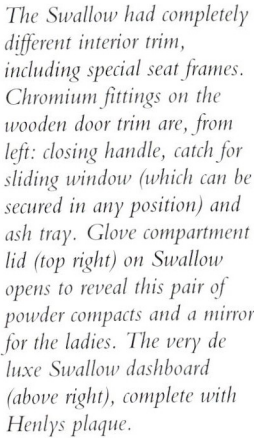

The Swallow had completely different interior trim, including special seat frames. Chromium fittings on the wooden door trim are, from left: closing handle, catch for sliding window (which can be secured in any position) and ash tray. Glove compartment lid (top right) on Swallow opens to reveal this pair of powder compacts and a mirror for the ladies. The very de luxe Swallow dashboard (above right), complete with Henlys plaque.

Swallow saloon was announced. This broke new stylistic ground and Henlys were very pleased to become agents for these chic new products, ordering several hundred cars – this liaison again carries through with Jaguar to the present day. The same style of saloon was built by Swallow, scaled up as necessary, on several other makes including Fiat, Wolseley and Standard. The latter created yet another link which formed the basis for Swallow's transformation first to SS and finally to Jaguar.

At first Swallows had an aluminium radiator shell but this was superseded by a nickel-plated steel one, and some cars apparently even left the works with stainless steel shells. Additional ventilation requirements were taken care of with the options of either a sliding sunroof or a pair of nautical pattern ventilators mounted on the scuttle – customers, in fact, often chose both.

Two years later, in time for the 1930 Motor Show, Mk2 versions of both the saloon and sports were announced. To the casual observer both looked very much as before from the side, but they had fresh frontal treatment. The radiator shell was less protruberent, had a central division and was chromium-plated, the false dumb iron valance was fully faired with twin sets of louvres, and a double-blade bumper was fitted with matching quarter bumpers at the rear. Startling colours had always been associated with these little cars but with the new model

Swallow excelled itself, offering many options from violet and cream to black and apple green.

The demise of the Austin Swallow during the summer of 1932 came not as a result of it falling from grace but as a direct result of its success, for William Lyons set up as a manufacturer in his own right with his first car, the SS1.

These three firms – Gordon England, Mulliner and Swallow – were easily the largest producers of bespoke coachwork on the Seven chassis, but many lesser outfits were quick to jump on the bandwagon. 'Lesser' refers to their size and the quantities produced, although sadly some of the products must have been viewed with abject horror by the hierarchy at the Austin factory and probably proved a rather bad joke to their owners who, to save face, had to carry on the pretence of having chosen something rather special. To be fair, many firms did offer a product that was both workmanlike and pleasing to the eye.

Some firms, such as Martin Walter of Folkestone, are known to have made but one body for a Seven. This could occur for a variety of reasons such as a special order or prototype but the following are some of the better known firms which were engaged at some time in the building of bodywork for the Seven. Their offerings, of varying quality, are described in alphabetical rather than chronological order, starting overleaf.

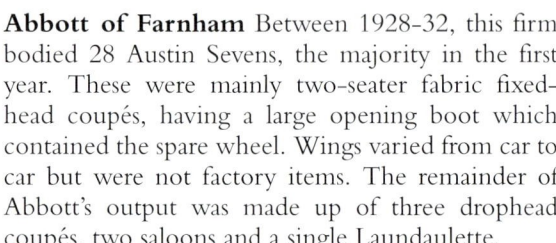

Swallow rear seat (far left) looks inviting but in truth is rather cramped; the rear blind is operated by the driver via a cord. A Pytchley sun roof badge (left, above), equally applicable to other Sevens fitted with one of these roofs. The threshold plate (left, below) is just one of numerous reminders of the Swallow identity.

Abbott of Farnham Between 1928-32, this firm bodied 28 Austin Sevens, the majority in the first year. These were mainly two-seater fabric fixed-head coupés, having a large opening boot which contained the spare wheel. Wings varied from car to car but were not factory items. The remainder of Abbott's output was made up of three drophead coupés, two saloons and a single Laundaulette.

AEW The North London firm of A E Wright began to produce its range of AEW Sevens at the beginning of 1933, by which time the practice of building alternative bodies for the Seven had practically ceased. AEW was obviously not a firm which looked to its elders or betters for inspiration as its products were fairly hideous. Nastiest was probably the two-seater sports: what the makers saw as a rakish sports car with louvred bonnet, lower body sides, cutaway doors and a fold-flat windscreen was rendered somewhat peculiar by a lengthy, angular tail with a flat top. This ridiculous motor car was further spoiled by bogus knock-on hub nuts, a windscreen-mounted spotlamp and forward-mounted headlamps. A four-seater version of the same device looked a little better but in common with some other small tourers of the period, such as the MG PA, it had too much overhang at the rear so any adventurous cornering with a full complement of passengers would be a sobering experience.

AEW's best attempt, on the 65 chassis, was brought out for 1934. It was called the AEW Speed 65 and the lines of the first two-seater had been cleaned up with the tail abbreviated, small scuttle humps added and the cycle wings replaced by quite passable swept ones.

AEW's swansong, the Z1, came in 1935. This looked the same as its stablemate, the K1 (the title now given to the Speed 65 version), apart from an MG-like chromium-plated radiator shell, the badge proudly proclaiming it to be an AEW. The Z1 was supercharged which prompted Mr Wright to claim a top speed of 80mph in his advertisements, which soon ceased due to his firm going out of business. Perhaps 200 cars were produced in total.

Alpe and Saunders This company, based at Kew, Surrey, is not remembered for its attractive coachwork and the few fabric coupés that it produced on the Seven chassis around 1928 were fairly staid in appearance. They should not be confused with the products of Mulliner, although they have Mulliner-type circular ventilators on the scuttle sides. The Alpe and Saunders coupés had small windows in the fixed-head to the rear of the doors.

ARC The ARC Manufacturing Company of Manchester planned to make a good number of what they described as a 'Coupette'. Its perpendicular-sided bodywork led into a nondescript rounded tail. The general lack of refinement of the 'Coupette' did not find many customers so the project was quietly abandoned.

Armstrong Starting in Goldhawk Road, Armstrong and Co operated from various addresses in the Shepherds Bush area of London. It produced a small number of drophead coupés on the Seven chassis. These had a high door line with winding windows, not unlike some of Windover's work, and the rounded tail was even higher giving them a hunchback appearance whether the hood was up or down. Eager to profit from any opportunity in the market, this firm also offered to convert tourers to saloons and modernise fabric saloons – by now falling from

Special Coachwork

A nice touch is the Swallow/Wilmot Breeden badge on the front bumper. Where the word Swallow appeared on the little powder compacts or the steering wheel boss, the 'W' was in the form of a bird in flight.

One of Arrow's fairly pleasing offerings, from an advertisement of 1933.

This sidelight was peculiar to the Swallow. Manufactured by Raydyot, it is now extremely rare.

fashion – by giving them new panelling in metal. It also acquired some unused stock from Gordon England and marketed new Cup and Stadium bodies in the early 1930s.

Arrow A version initially created in 1929 as alternative sporting coachwork for the Austin Seven, with the bodies made by A P Crompton of Merton, south London. This title eventually referred also to similar styles erected upon Wolseley Hornet and Austin Ten chassis. The first cars were fabric open two-seaters with sloping tails and swept wings devoid of running boards. At first the spare wheel was carried inside the rear but later it was mounted externally. At various stages during the first two or three years different bodies were experimented with, and a few coupés and pointed-tail two-seaters were produced: in 1930 the firm moved to Hanwell.

During 1931 A P Crompton started to make quite a pleasant little four-seater tourer with cut-down doors and the same wings as the earlier two-seater. In short-chassis form it looked a little stunted and accommodation in the back must have been crushed, but when lengthened and built on the longer chassis it acquired good proportions.

In 1933, by which time H A Saunders of Golders Green had taken over the marketing of these cars from Normand Garages, the Arrow 65 was introduced. This, rather like the AEW, was designed to appeal to the relatively impecunious 'Promenade Percy' rather than the serious sports car enthusiast. A travesty of a swept-wing J2 MG, it almost groaned under the weight of headlamp stone guards, knock-on hub caps, special filler caps, racing bonnet strap and other goodies, but at least the heat build-up from the massively powerful engine could escape through the multiplicity of louvres that covered the bonnet! There must have been a goodly supply of Percys who wished to promenade in these things because they were made for two years; some of the bodywork on Arrows was executed by Wittingham and Mitchell during this period. The 75 version that appeared in 1935 looked the same but was fitted with a Speedy engine – and for the real speed merchant a supercharged version was available.

Avon In later years this firm, actually the new Avon Body Company of Warwick, was well-known for the Standard Avon, a competitor of the Standard Swallow. Avon also produced bodies, often sports coupés, for such firms as Lanchester during the early 1930s, but some of its early work was done on the Austin Seven chassis. In 1927 Avon started to produce quite a pretty little fabric-covered open two-seater with a rounded tail and an attractive downward curve to the rear deck. It declined to use Austin's mudguards and came up with rather more flowing examples of its own. In common with many other firms, the running boards were aluminium-covered on Avon's offering.

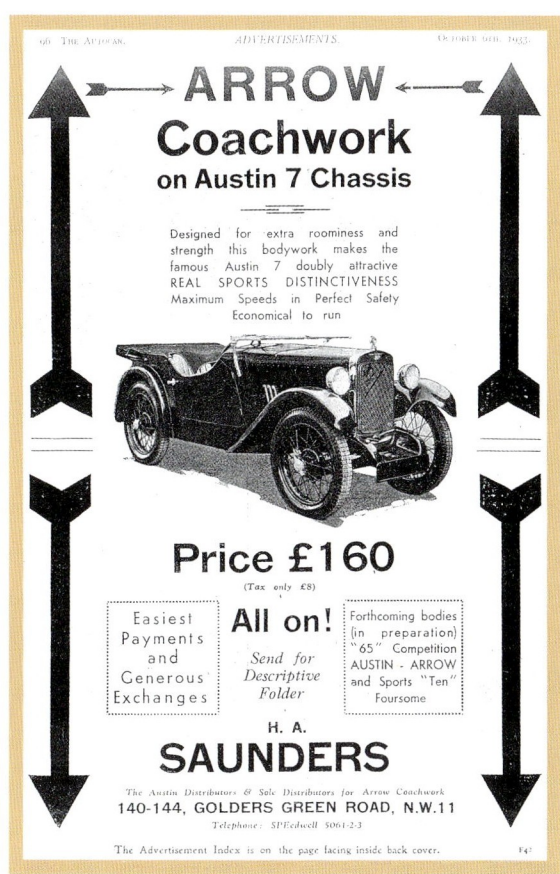

A year later a little fabric coupé, called the Swan, appeared. This had a fold-back sunshine roof and less flowing wings without running boards. The bonnet was louvred, as were other bodies made for the Seven by this company, before the parent factory introduced them on its own bodies. Avon's last style for the Seven was introduced in the summer of 1929. This was an open two-seater rather in the form of the Morris Minor two-seater, but with helmet type front wings similar to those used on the Standard Avon of the time.

Boyd Carpenter For some time this firm, whose works were in Kilburn, north London, specialised in tuning Austin Sevens after its founder, Boyd-Carpenter, left his job with Gordon England. In addition to producing a very racy body, this firm considerably tuned the engines – including often fitting LAP overhead valve cylinder heads – and lowered the suspension, among other modifications. The bodywork had a long pointed tail, cycle wings, dumb iron fairing, louvred bonnet and side valances. Although this car exhibited some of the fashionable boy-racer characteristics, it was in fact a workman-like and well-conceived little sports car. Quite a number were made during the late 1920s.

Burghley This was the first firm which successfully produced specialised coachwork for the Austin Seven in any quantity. Produced in London by Wilson Motors of Victoria, the two Burghley

123

models appeared during 1924. The design of the open two-seater sports owed more to Paris than London, having a pointed tail with hardwood-planked deck, diminutive doors and a 'V' windscreen. Wing design altered from car to car, some even affecting the French clamshell type, while others had blade wings front and rear sloping down to a tiny running board in the centre. The other model was a Saloon Laundalette, a much more conventional product but still popular. Both versions were made for several years.

Cole and Shuttleworth The antecedents of this firm were the Cole family which practised around Hammersmith prior to World War I, following which a partnership was formed and this business then operated from the Kings Road in Chelsea. Its venture with the Seven was a boat-tailed two-seater, panelled in aluminium with helmet wings and, like many two-seater Austins with non-standard coachwork, a small hatch in the rear deck. Surviving pictures do not show the presence of a spare wheel, a necessary encumbrance but hard to accommodate with this type of bodywork. The instrumentation, upholstery and general finish on these cars was of a better standard than some other custom bodies fitted to the baby Austin. Production took place from 1926 to 1928.

Duple Much better known for its bus and lorry bodies, this company from Hendon made car bodies at various times. Its first attempt with the Seven, in 1926, was a semi-commercial version which could take on the form of a van or tourer by the substitution of various body sections, seats and hood. It was over-ingenious and found few buyers. A rather attractive two-seater went into production the following year. The problem of the spare wheel was solved by mounting it underneath the shapely little tail, in the same fashion as the duck's back 12/50 Alvis which this body somewhat resembled.

Hawk In 1930 the Matchless Motorcycle factory, situated in south-east London, came to the conclusion that as the Austin Seven had taken away a good deal of its sidecar trade, it might as well try and profit by this. Accordingly it designed and started building upon the Seven chassis fabric-covered, pointed-tail sporting bodies, which bore more than just a passing resemblance to an MG M type. Matchless even manufactured its own radiator shell complete with a Hawk badge, the shell being just different enough from the MG not to cause comment. Below this, rather spoiling the effect, was a wide pressed-steel valance, carrying a large rectangular number plate which covered the front axle and suspension. The cycle wings differed from Abingdon's baby in that they were fully valanced at the front and accordingly had to turn with the steering. An articulated crossbar which carried the headlamps was also incorporated into this set-up. Some bright spark at Matchless may have thought that this scheme would point the lamps around corners. Of course, due to the geometry, they would forever point straight ahead. Not surprisingly the Hawk was unable to compete with the M-type and after a while Matchless abandoned the idea and went back to doing what it was best at.

Hoyal Normally this firm, just around the corner from Brooklands at Weybridge, was associated with bodywork for Daimler cars and suchlike. Around 1928, to keep the workforce employed during lulls in orders, it built several fabric saloons for the Austin Seven. It then put into limited production a rather unlovely two-seater with large luggage boot, cycle wings and louvred side valances. This model lasted until 1930.

Hughes Along with the Burghley and the fabled Gruzelier, the only evidence of which is an advertisement for one on sale in 1925, the bodies made by sidecar manufacturer Thomas Hughes of Birmingham are the earliest by an outside maker. This car was styled in the Alvis 12/50 duck's back idiom with the spare wheel mounted under the tail. The body used no Austin panels and was without doors. It had a 'V' screen and a set of rather pretty wings, which had high running boards and carried nickel-plated headlamps. This delicate looking little car was altogether very pleasing to the eye.

Wilson Motors produced quite a number of its 'Burghley' Sports and Saloon Landaulette models during the mid-1920s.

Special Coachwork

Jarvis was among the better firms building bodies for the Seven. Headlamps are mounted in the same way as Gordon England cars, so perhaps the necessary components were bought from just around the corner in Putney.

Jarvis The bulk of this firm's output over the years was fairly skimpy sporting bodywork on a variety of cars from Bugatti to Wolseley Hornet, so it was not surprising that it turned its hand to the Austin Seven. In the late 1920s it built a few open two-seaters, fabric-covered and with pointed tails.

KC Yet another small firm, KC Bodies of New Kings Road, Chelsea, saw a little sports car on the Austin Seven chassis as a profitable exercise. Brought out in 1928, it was better contrived than some but played safe with its styling, which followed the boat-tailed fashion. Fully valanced helmet wings of a good flowing shape were fitted. At first the rear deck had a small hatch but by 1929 the rear section hinged up. The bonnet was louvred, a 'V' screen was used, and unusually the body panels curved inwards at the base to form an undertray in true racing fashion. The same firm undertook tuning on Sevens and it was possible to order your KC Sports with a hotter motor if the fancy took you.

Taylor Just around the corner from KC Motors, in South Kensington, H Taylor and Company produced its idea of an Austin Seven sports two-seater which, 'V' windscreen apart, differed completely from its near-neighbour's products. A prim, curved tail carrying the spare wheel and rakish wings, with the wheels often covered by polished aluminium wheel discs, ensured adequate sales for this to be made for two or three years, commencing in 1927.

Wydoor This was the name given by G Wylder and Company of Kew, Surrey, to a range of fabric bodies it built for the Austin Sixteen, Twelve and Seven during the late 1920s. That fitted to the Seven had, as the name suggests, wide doors which extended partially over the rear wheel-arches and which, in common with other such layouts, necessitated sliding windows. From the rear they are recognisable on account of the inverted D-shaped rear window and, in common with the Mulliner saloon, they have false landau irons, although these are longer on account of there being no rear side windows. Early examples had round scuttle ventilators which were changed for rectangular ones by 1928. Factory wings were used.

In addition to the above firms and others which made perhaps only one or two cars, Pytchley Autocar, which supplied sunroofs to Austin and many other manufacturers, produced in around 1928 a few Austin Seven fabric saloons, complete with its own products.

Tickford also executed several conversions on saloons during the early 1930s, equipping them with its patented folding top. This operated by inserting a winding handle into an escutcheon on the side of the rear bodywork.

DATA SECTION

IDENTIFICATION & DATING

In addition to any identification plates, all Austin Sevens had both chassis and engine stamped with their serial numbers. The chassis number is found on the top of the nearside chassis rail between the engine mounting points. The engine number, prefixed by the letter M, is on the nearside of the crankcase, just below and to the front of the tappet cover.

At first there were no separate chassis and car numbers, and production began in March 1923 at number A1-101. Around A1-2650 was reached by the end of 1923, and around A1-7800 by the end of 1924. During the early summer of 1925 car numbers moved on to the A2 series, and from this point until 1936 chassis and car numbers were dissimilar; chassis numbers reached perhaps 15000 by the end of 1925.

In 1926, with a change to A3 car numbers in May, chassis numbers advanced to about 29000. During 1927, before A5 car numbers were introduced in the autumn, the A4 series was made and it has been suggested that this was reserved for cars supplied to outside coachbuilders. Many, but not all, Mulliner and Gordon England cars do carry this number, but in addition there are many standard-bodied cars in the A4 series. Chassis numbers reached just over 50000 by the end of 1927.

In 1928 two car number series were introduced, first A6 early in the year and then A7 at the end of the summer; chassis numbers reached nearly 75000 that year. January 1929 saw the A8 series and then during the summer the A9 series, while chassis numbers ran through to just over 100000 by the close of the year, by which time B series car numbers had been embarked upon, the first range without a suffix to the B. During 1930, the last year of the vintage period, chassis numbers reached about 125000, and series B1 and B2 car numbers were begun in June and September respectively.

The B3 series started in March 1931 and then the B4 during the summer, while chassis numbers ran on to about 145000 by the end of the year. The next year chassis numbers reached about 166000 and car number series B5 and B6 were made; 1933 was another year with production of around 20,000 vehicles and chassis numbers went through to nearly 187000.

The Ruby was introduced at chassis number 198596 during July 1934. Just before this, with B series car numbers having run through to B9, the C series arrived at the beginning of June but it was to be short-lived as a new car number system was introduced in August - the Ruby was ARQ, the tourer AAK and the Pearl AC. Chassis numbers reached just over 210000 that year.

Seven production peaked during 1935, some 27,000 cars leaving the factory, and consequently chassis numbers arrived at around 237000 by the end of the year. At this time car numbers were prefixed AVJ for vans and AEB for Sports models.

In July 1936 the factory abandoned the system of separate car and chassis numbers, the two henceforth becoming the same. Chassis numbers reached nearly 260000 at the end of 1936, about 282000 for 1937, and just under 290000 for 1938. Vastly reduced output during the first few months of 1939 only advanced the numbers to 291000, at which point production ceased.

Engine numbers do not match chassis numbers, generally running a few hundred ahead. The gap widened as the years passed, with a 1937 engine number, for example, typically being 1200-1400 ahead of the chassis number. It seems likely that this situation was caused by selected engines being used for experimental projects or as replacement units for customers. Ulster models, however, are an exception to this general rule, for their engines normally have lower numbers because they were specially built in batches and then held in stock, sometimes for as long as a year.

The serial which designated body type and number was stamped on the transmission tunnel, normally to the rear of the handbrake aperture.

Any apparent discrepancies between the above information and the production figures quoted (right) are due to the latter often only being counted until the autumn of a particular year. For instance, production figures for 1923 run only until October, so cars built in November and December are counted in the 1924 figures.

Unfortunately, full factory records do not exist, but some of the original build ledgers from 1929-31 have survived. If you own a car of this period, you may be able to find out full details of its original specification by contacting the British Motor Industry Heritage Trust and quoting your chassis number.

PRODUCTION FIGURES

Year	Total	Model breakdown
1923	1936	Until October
1924	4700	From November 1923
1925	7043	
1926	14,000	
1927	22,500	
1928	22,709	Tourers 8233, saloons 2142, fabric saloons 4331, coupés 51, vans 316, Mulliner saloons 2868, Gordon England saloons 1024, Gordon England Cups 908, Gordon England vans 101
1929	26,447	Tourers 5304, saloons 3607, fabric saloons 8309, two-seaters 107, coupés 347, vans 392, military 118, Mulliner saloons 2549, Gordon England saloons 683, Gordon England Cups 494, Gordon England vans 104
1930	23,826	Tourers 3187, saloons 11,101, fabric saloons 5120, coupés 128, vans 746, sports 48, military 40, Mulliner saloons 1269, Gordon England saloons 178, Gordon England milk vans 56
1931	21,282	Tourers 1647, saloons 12,758, de luxe saloons 2970, fabric saloons 782, two-seaters 559, coupés 7, vans 1194, sports 97, Mulliner saloons 424, Mulliner vans 92
1932	20,121	Tourers 1062, saloons 2998, de luxe saloons 12,434, two-seaters 476, vans 2364, sports 22, military saloons 127, Mulliner vans 18
1933	20,475	Tourers 1484, saloons 2765, de luxe saloons 11,837, two-seaters 865, vans 2497, sports 234, military two-seaters 4
1934	22,542	Tourers 1449, saloons 2901, de luxe saloons 13,638, cabriolets 76, two-seaters 1015, vans 2380, sports 228, military two-seaters 141
1935	27,280	Tourers 1362, saloons 4343, de luxe

Early types of body number plate, seen on a 1923 AB tourer (above) and a 1926 R saloon (below). Stewart speedometer and petrol tank dipper hanging on the hook are accessories on the earlier car; note that the trim panels are secured by tacks.

Just discernible at top left, body number was stamped on the transmission tunnel, in this case AF365. Also worth noting in this picture of a 1930 AF tourer are the screws securing the forward portion of the transmission tunnel, and the starter button. Besides chassis and car number plates, the supercharged Ulster (below) has an instruction plate for the supercharger; the rear of the air pump is in the foreground.

Sir Herbert Austin was keen to protect his numerous patents relating to the Seven and from fairly early on each car prominently displayed a plate such as this. This one is on a 1937 ARR Ruby and would be different, for example, from that on a 1927 saloon.

By the time this 1937 ARR Ruby was made, body and chassis numbers were the same; this amalgamation took place at chassis number 249701 in July 1936.

1936	23,500
1937	23,000
1938	8500
1939	1000

saloons 15,864, cabriolets 1052, two-seaters 735, vans 1916, sports 220, military two-seaters 396

In the absence of complete chassis records having survived, the figures compiled by R J Wyatt nearly 30 years ago are quoted. These are, in his own admission, incomplete and in some cases approximate, but in all probability they represent a fairly accurate picture of events.

Export figures are available for the following years: 1928, 4871; 1929, 7669; 1930, 4619; 1931, 3656; 1932, 3966; 1933, 3372; 1934, 3740; 1935, 6536.

COLOUR SCHEMES

As far as can be determined from examination of available original factory literature, the following is a list of standard colour schemes used on the Seven. Any colours in the range could be specified for a model not supplied in that finish, but notification had to be given in advance. In addition colours not in the Austin range could be had for an extra charge. Special colours were sometimes used on cars destined for exhibitions.

1923-26
Tourer Kingfisher Blue, Elephant Grey, Primrose, Dark Royal Blue, Maroon, Auto Brown **Sports** Kingfisher Blue **Van** Supplied in grey primer and painted in a stock colour at extra charge.

1926-27
Tourer & saloon Maroon, Oxford Blue, Kingfisher Blue, Primrose, Dove Grey, Dark Green.

1928
Tourer & saloon Maroon, Auto Brown, Primrose, Black, Royal Blue, Battleship Grey, Maroon (two-tone), Blue (two-tone), Light Brown and Grey, Brown and Tan.

1929
Tourer Putty (all other colours as for 1928).

1930
Tourer & saloon Opal Blue, Primrose, Blue, Beige and Maroon, Auto Brown, Maroon, Black, Battleship Grey **Note** To give an idea of some of the non-standard colours in which cars left the factory, the following are taken at random from the factory ledger for this year: Cambridge Blue and Turquoise, Blue and Electric Blue, Maroon and Silver Grey with Black leather (saloon), Almond Green, Khaki and Brown, Silver Grey (tourer), Brown and Cream, Blue and Grey with Cerise leather (coupé), Beige and Maroon, Maroon and Grey (coupé).

1931
Tourer & saloon Royal Blue, Maroon, Opal Blue, Maroon and Black, Royal Blue and Black, Brown and Black, Green and Black.

1932
De luxe saloon Light Blue/Black top with Blue coachlines, Light Auto Brown/Black top with Cream coachlines, Light Maroon/Black top with Red coachlines, Opal Blue/Black top with Cream coachlines, Fawn/Black top with Cream coachlines **Saloon** Light Royal Blue with Cobalt Blue band, Light Maroon with Beige band **Tourer & two-seater** Light Royal Blue, Light Maroon, Opal Blue.

1932-33
De luxe saloon Royal Blue/Black top with Blue coachlines, Maroon/Black top with Red coachlines, Westminster Green/Black top with Cream coachlines, Black, Auto Brown/Black top with Cream coachlines, Dual Beaver with Blue coachlines **Saloon** Royal Blue with Blue band, Maroon with Beige band, Black with Brown band **Tourer** Royal Blue with Black band, Maroon with Black band, Auto Brown with Black band, Opal Blue with Black band **Two-seater** Royal Blue with Black band, Fawn with Black band.

1933-34
De luxe saloon Royal Blue/Black top with blue coachlines, Maroon/Black top with Red coachlines, Westminster Green/Black top with Cream coachlines, Black with Red, Green or Brown coachlines, Turquoise Blue/Black top with Cream coachlines, Dove Grey/Black top with Cream coachlines, Pueblo Brown/Malay Brown top with Yellow coachlines **Saloon** Black with Brown coachlines, Royal Blue with Blue coachlines, Maroon with Beige coachlines **Tourer** Royal Blue with Blue coachlines, Maroon with Red coachlines, Auto Brown with Cream coachlines **Two-seater** Royal Blue with Black band, Dove Grey with Black band **Note** Additional colours in the Austin range that could be ordered were Silver Grey and Cherry Red **65 & Nippy** Primrose, Turquoise Blue, Cherry Red.

1934-35
Ruby & Pearl Royal Blue with Blue coachlines, Dove Grey with Cream coachlines, Westminster Green with Cream coachlines, Black with Red, Green or Brown coachlines **Ruby only** Maroon with Red coachlines **Pearl only** Cherry with Red coachlines **Ruby fixed-head** Royal Blue with Blue coachlines, Maroon with Beige coachlines, Black with Brown coachlines **Tourer** Royal Blue with Light Blue band, Maroon with Black band, Auto Brown with Black band **Opal** Royal Blue with Light Blue band, Dove Grey with Black band **Nippy & Speedy** Primrose, Turquoise Blue, Cherry Red, Black.

1935
Ruby & Pearl Royal Blue with Light Blue coachlines, Westminster Green with Green coachlines, Black with Brown, Red or Green coachlines, Dove Grey with Gold coachlines **Pearl only** Cherry Red with Red coachlines **Ruby only** Maroon with Red coachlines **Ruby fixed-head** Royal Blue

It is most unusual to come across a car with all its correct literature, including sales brochure. Many amusing little mascots, such as the irate cat on this 1926 R saloon, were available for Seven owners.

The factory ran vigorous advertising campaigns in most of the popular motoring magazines of the day.

Some of the many publications from the factory, in chronological order (with factory publication numbers in brackets): 'The Austin Magazine and Advocate' of October 1927, List of Spare Parts of January 1929 (670), 'Life' ('As you like it with an Austin Seven') of 1929 (701), Sports Model of 1931 (730C), 'Give Way to Happiness' postcard of 1931, Lubrication chart of April 1934, List of Spare Parts of March 1935 (1218A) and sales leaflet from July 1938 (1677/1) that rather naughtily states that over 300,000 Sevens had been sold.

Original Austin Seven

with Blue coachlines, Maroon with Beige coachlines, Black with Brown coachlines **Tourer** Royal Blue with Blue band, Maroon with Black band, Auto Brown with Black band **Opal** Royal Blue with Light Blue band, Dove Grey with Black band **Nippy & Speedy** Primrose, Turquoise Blue, Cherry Red.

1936
Ruby & Pearl Royal Blue with Light Blue coachlines, Westminster Green with Green coachlines, Black with coachlines to match upholstery, Dove Grey with Gold coachlines **Pearl only** Cherry Red with Red coachlines **Ruby only** Maroon with Red coachlines **Ruby fixed-head** Royal Blue with Light Blue coachlines, Maroon with Red coachlines, Black **Tourer** Royal Blue with Light Blue coachlines, Maroon with Red coachlines, Auto Brown with Tan coachlines **Opal** Royal Blue with Light Blue coachlines, Dove Grey with Gold coachlines **Nippy** Primrose, Turquoise Blue, Cherry Red, Black.

1936 (from August)
Ruby & Pearl Royal Blue with Blue coachlines, Westminster Green with Green coachlines **Ruby** Maroon with Red coachlines, Ash Grey with Grey coachlines, Black with coachlines to match upholstery **Pearl** Cherry Red with Red coachlines, Black with Fawn coachlines **Ruby fixed-head** Royal Blue with Blue coachlines, Maroon with Red coachlines, Black with Fawn coachlines **Tourer & Opal** Royal Blue with Blue coachlines, Ash Grey with Grey coachlines, Auto Brown with Tan coachlines **Nippy** Primrose, Turquoise blue, Cherry Red, Black.

1936 (from November)
Ruby & Pearl Royal Blue with Blue coachlines, Westminster Green with Green coachlines, Ash Grey with Grey coachlines, Polychromatic Blue with Blue or Grey coachlines, Polychromatic Grey-Green with Grey coachlines **Ruby** Maroon with Red coachlines, Black with coachlines to match upholstery **Pearl** Cherry Red with Red coachlines, Black with Fawn coachlines **Ruby fixed-head** Royal Blue with Blue coachlines, Maroon with Red coachlines, Black with Fawn coachlines **Tourer & Opal** Royal Blue with Blue coachlines, Ash Grey with Grey coachlines, Maroon with Red coachlines **Nippy** Primrose, Turquoise Blue, Cherry Red, Black.

1937 (from February)
As for November 1936, with the exception that the Ruby and Pearl cars finished in Polychromatic Blue now only had Blue coachlines.

1937 (from November)
As for February 1937, apart from the Tourer and Opal which now could also be had in Auto Brown. During this period the van version continued to be supplied in grey primer but could be finished in Red, Blue, Yellow, Green or Black at extra cost.

1938
Ruby & Pearl Royal Blue, Princess Blue, Ash Grey, Black **Ruby** Maroon, Bluebird Blue **Ruby fixed-head** Royal Blue, Maroon, Black **Tourer & Opal** Royal Blue, Ash Grey, Maroon.

EXTRAS & ACCESSORIES

The factory did not offer a range of extras for the Seven but at various times during the life of the car the factory listed alternative equipment: for instance 16in road wheels and 4.75 Extra Low Pressure tyres were offered for the Nippy or Speedy. Such options are mentioned in the relevant sections of this book.

It quite soon became obvious, however, that there was a gap in the market and this was quickly filled by the various specialist accessory manufacturers. Although their wares are not strictly speaking within the scope of this book, I have included a list of some that may be encountered. Apart from the fun that they can provide, some are of very real practical use and some, for example the David Harcourt oil pressure gauge, were later adopted by the factory. This is not intended to be a complete list of every gadget made for the Seven and they are either referred to by their maker's or trade name.

Powell and Hanmer External mirror 244 or 244X; horn 277 exits from side of scuttle, also for Mulliner-bodied cars; horn 276 fits under bonnet; step mat 356 or 356X; radiator muff A1; spare petrol can and carrier 309, sloping base, fits on front wing.

Stadium Gear lever extension; baby spot light; baby fog light; spare petrol can with dipstick; de luxe step mat; Flexifit horn, fits under bonnet and clipped to steering column; horn, fits through scuttle; junior autoscope mirror; door lock and handle; horn ring, 1927/28 models; radiator thermometer; rear blind, for saloons; roof net, for saloons.

Desmo Running board scraper/mat; flexi cover for propshaft universal joint; radiator tie bar, fits under headlamps on post-1928 models.

Wilmot Breeden Door lock set; bumpers; 'Prince' calormeter; radiator shutters (in fact a blind operated by cable and winder).

Duco Radiator muff; rubber step mat; eight-day 'midget' clock.

Prima Petrol dipstick; oil gauge dipstick; combined petrol and oil can; petrol can carrier.

Clayrite Horn No 35, through-scuttle mounting; mirror No 36.

Easyfit Luggage carriers, a very comprehensive range: 8B, first type; 8C, 124/5; 8D, 1926 with wing stays; 8E, 1926 without wing stays; 8F, 1927/29 tourer, saloon and 1928/29 fabric saloon; 8G, 1927/28 Gordon England saloon; 8H, 1928/29 Mulliner saloon; 8K 1929 Gordon England saloon; 8L, 1930/31 saloon and tourer, etc; pressed steel variant with the prefix SE (eg, SE 8L) was available for cars made after 1928; bumpers.

Ashby Easy-fit bumpers.

Autovac Petrol gauge.

Bodelo Four-wheel brake attachment, for cars without uncoupled brakes.

Wefco Spring gaiters.

Ewart Wheel discs, steel or aluminium.

Climax Stainless steel bumpers.

Wilcot Seat covers, striped fabric.

New Century Bumpers.

Auster Radiator muff; windscreen visor.

Benjamin Radiator shutters; these were particularly impressive.

Enots 'Minor' hydraulic jack.

David Harcourt 'Linkula' oil gauge, 0-20lb to replace oil button.

Fabram Radiator muff; spring gaiters.

Midland Car mats; radiator muff, various types with even a special version (1503) for the Swallow.

Smiths Radiator shutters, either hand or thermostatic operation.

Olympic Baby spotlight, no 1225.

Hutchinson Pneumatic seats, to replace cushions at the front.

Frikke Steering damper.

B&D Steering stabliser.

Brooklands Steering wheel; radiator stone guard.

BB Spare petrol can, fits under seat.

The above is only a part of an enormous range available which also included special oil fillers and plates to facilitate clutch race oiling. Added to all these was a plethora of other wares that were not specifically for the Seven but made driving more fun. How about a Cozytowz foot warmer or a lighthouse mascot which doubled as a parking lamp and shone green or mauve?

Through-scuttle horn on 1926 R saloon is by Desmo, petrol can by P&H (Powell & Hanmer).

The tax disc holder is by Raydyot and the petrol can by P&H (Powell and Hanmer); the petrol can was designed when the trailing section of an Austin Seven front wing was straighter than on this AE tourer.

External mirror on 1930 AE tourer is by Desmo, through-scuttle horn by an unidentified maker, fire extinguisher by Pyrene, and foot scraper by Stadium.

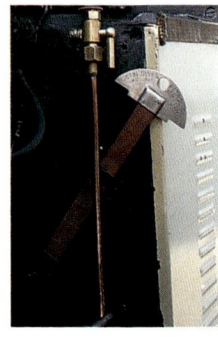

A very sensible accessory, before the cars came equipped with a petrol dipstick, was a 'Prima' petrol gauge. This one is clipped under the bonnet of a Swallow saloon; note also construction of bulkhead sides.